Fundamentals of AutoCAD®

Prentice Hall Modular Series for Engineering

Now, you can tailor your course materials to satisfy the needs of you and your students. With the Prentice Hall Modular Series for Engineering, you can mix and match these concise, inexpensive books for your classes. They are ideal for courses in which a variety of languages and software are being covered. All books focus on using real-world applications to help motivate your course.

Current Modules Include:

Introduction to MATLAB for Engineers and Scientists
Delores Etter–University of Colorado, Boulder
1996, 145 pp. 0-13-519703-1

Introduction to C++ for Engineers and Scientists
Delores Etter–University of Colorado, Boulder
1997, 160 pp. 0-13-254731-7

Introduction to ANSI C for Engineers and Scientists
Delores Etter–University of Colorado, Boulder
1996, 164 pp. 0-13-241381-7

Introduction to Fortran 90 for Engineers and Scientists
Larry Nyhoff and Sanford Leestma–both of Calvin College
1997, 336 pp. 0-13-505215-7

Fundamentals of AutoCAD
Mark Dix and Paul Riley–CAD Support Associates
1998, 250 pp. 0-13-860362-6

Engineering Design: A Day in the Life of Four Engineers
Mark Horenstein–Boston University
1998, 150 pp. 0-13-8660242-8

Introduction to the Internet for Engineers and Scientists
Scott James–GMI Institute
1998, 150 pp. 0-13-856691-7

Introduction to ProENGINEER
Jeffrey Freeman and Andrew Whelan–both of the University of Iowa
1998, 150 pp. 0-13-861048-7

Future Modules Include:

Engineering Ethics
Engineering Problem-Solving
Introduction to MS Word
Introduction to WordPerfect
Introduction to Excel
Introduction to Lotus 1-2-3

Fundamentals of AutoCAD®

Mark Dix and Paul Riley

CAD Support Associates

Prentice Hall
Upper Saddle River, New Jersey 07458

Library of Congress Cataloging-in-Publication Data

Dix, Mark
Fundamentals of AutoCAD / Mark Dix and Paul Riley
p. cm. -- (Prentice Hall modular series for engineering)
Includes index.
ISBN: 0-13-860362-6
1. Computer graphics. 2. AutoCAD (Computer file). I. Riley, Paul.
II. Title. III. Series.
T385.D6325 1998
620' .00285'5368--dc21 97-28518
CIP

Editor-in-chief: **MARCIA HORTON**
Acquisitions editor: **ERIC SVENDSEN**
Director of production and manufacturing: **DAVID W. RICCARDI**
Managing editor: **BAYANI MENDOZA DE LEON**
Cover director: **JAYNE CONTE**
Production editor: **IRWIN ZUCKER**
Copy editor: **SHARYN VITRANO**
Manufacturing buyer: **JULIA MEEHAN**
Editorial assistant: **ANDREA AU**

Printed in the United States of America

10 9 8 7 6 5 4 3 2 1

ISBN 0-13-860362-6

PRENTICE-HALL INTERNATIONAL (UK) LIMITED, *London*
PRENTICE-HALL OF AUSTRALIA PTY. LIMITED, *Sydney*
PRENTICE-HALL CANADA INC., *Toronto*
PRENTICE-HALL HISPANOAMERICANA, S.A., *Mexico*
PRENTICE-HALL OF INDIA PRIVATE LIMITED, *New Delhi*
PRENTICE-HALL OF JAPAN, INC., *Tokyo*
SIMON & SCHUSTER ASIA PTE., LTD., *Singapore*
EDITORA PRENTICE-HALL DO BRAZIL, LTDA., *Rio de Janeiro*

Drawing credits: Hearth Drawing courtesy of Thomas Casey: Double Bearing Assembly compliments of David Sumner, King Philip Technical Drawing; Isometric Drawing Flanged Coupling courtesy of Richard F. Ross; Isometric Drawing of Garage courtesy of Thomas Casey.

Contents

Chapter 3 54

Chapter 4 78

Chapter 5 102

Preface

This book is a shortened version of our book *Discovering AutoCAD Release 13 for Windows.* It uses the same format, the same concepts, and the same teaching methods. But while *Discovering AutoCAD* is designed for a full year of instruction and practice, this introduction includes less content and fewer drawing exercises so that it may be completed in a few weeks.

Developing expertise in drawing on a CAD system takes hours of practice. Like driving a car or playing a musical instrument, it cannot be learned by reading about it or watching someone else do it.

Accordingly, this book, like its predecessor, takes a very active approach to teaching AutoCAD®. It is designed as a teaching tool and a self-study guide and assumes that readers will have access to a microcomputer CAD workstation. It is organized around drawing exercises or tasks that offer the reader a demonstration of the commands and techniques being taught at every point, with illustrations that show exactly what to expect on the computer screen when steps are correctly completed. The focus here is on the beginning AutoCAD user who needs a brief but thorough introduction to AutoCAD drawing principles.

In this text we strive to present an optimal brief learning sequence. Topics are carefully grouped so that readers will progress logically through the AutoCAD command set. Explanations are straightforward and focus on what is relevant to actual drawing procedures. Most important, drawing exercises are included at the end of every chapter so that newly learned techniques are applied to practical drawing situations immediately. The level of difficulty increases steadily as skills are acquired through experience and practice.

Drawing exercises at the end of all chapters are reproduced in a large, clearly dimensioned format on each right-hand page with accompanying tips and suggestions on the left-hand page. Drawing suggestions offer time-saving tips and explanations on how to use new techniques in actual applications. While the focus is on mechanical drawings, there are also architectural drawings in many chapters.

We would like to thank the many people who have helped us in the preparation of this book. To begin with, thanks to Eric Svendsen, our editor at Prentice Hall, for his encouragement and support, and to Irwin Zucker, for his expert production editing.

We are grateful to the students, staff, and administration at Mount Ida College for their interest in this project and commitment to CAD education. We are grateful to a number of people at Autodesk for the use of software and support.

Finally, thanks to Lauri O-Brien at DTI Technologies for her expert advice and support.

Mark Dix and Paul Riley

Road Map

You will notice that all the chapters in this book follow the same layout. We have included this road map to help you find your way around. The following is a description of each of the major sections of the chapter format along with sample entries.

COMMANDS (SAMPLE)

HELP	MODIFY
HELP	ERASE
	MOVE
	ROTATE

The COMMANDS section lists the commands and topics that are introduced in the chapter. Headings are taken from the AutoCAD standard menu and may refer to pull down menu headings or toolbars. For example, you will see HELP under its own heading because there is a Help pull down menu; and ERASE, MOVE, and ROTATE under Modify, refering to the Modify toolbar.

OVERVIEW

Each chapter begins with a brief overview that gives you an idea of what you will be able to do with the new commands in that chapter.

TASKS (SAMPLE)

1. Read about introductory tasks.
2. Read about drawing projects and the HELP command.
3. Begin Chapter 1.

All the tasks that make up a chapter are listed at the beginning of the chapter. This tells you at a glance exactly what you will be required to do to complete the chapter.

Each chapter consists of two types of tasks. The body of the text includes fully illustrated, step-by-step exercises in which new commands and techniques are introduced. Then, at the end of the chapter, you will find drawing projects that require the use of the new commands and techniques. By completing the exercises in the chapter, you will learn the skills needed to complete the drawings at the end of the chapter.

TASK 1: Introductory Tasks

Each task includes specific instructions along with explanations, illustrations, and feedback about what will happen on your computer screen when you carry out the instructions. Typically, there will be an instruction followed by AutoCAD's response and any information we feel is necessary or helpful.

All instructions which require an action on your part are preceded by an arrow ">".

The HELP Command

The following sample instructions show how to use AutoCAD's HELP command. It is included here for reference and as an example and is not necessarily intended for actual execution at this time. The HELP feature is useful, however, and you are encouraged to consult it frequently as you progress through the book.

(*sample instructions*)

> Select "Help" from the pull down menu bar at the top of the drawing editor.

This will cause a menu to drop down below the word Help. You can also enter HELP from the Standard toolbar or the command line, but the results will be somewhat different from what is described here.

> Select "Search for Help On . . ." from the Help menu.

This will call up the AutoCAD Help Search dialogue box shown in Figure 1.

To move quickly through the list, you can click in the edit box and type the name of a command and press enter. Alternativley, you can move the cursor arrow over the scroll bar arrow and scroll down until you find the item you are looking for. Then move the cursor over the item and press the pick button twice to highlight it and to highlight the "Go To" box. Let's try it.

> Type "line".

Notice that AutoCAD responds to each keystroke so that you may not have to type the whole word.

> Press enter.

"LINE Command" will be shown as the selected topic in the lower box.

> Press enter or click on "Go To" to call up information on the selected command or topic, in this case the LINE command.

This will call up information on the LINE command as shown in Figure 2.

> Click on File and then "Exit" to exit the HELP command.

(*end of sample*)

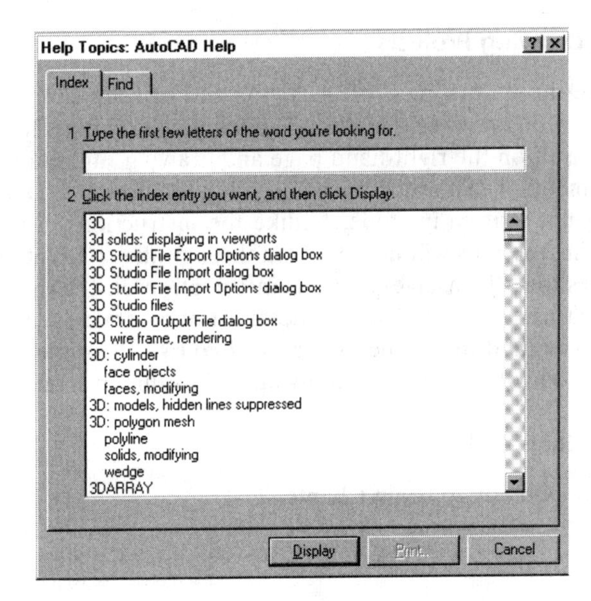

Figure 1

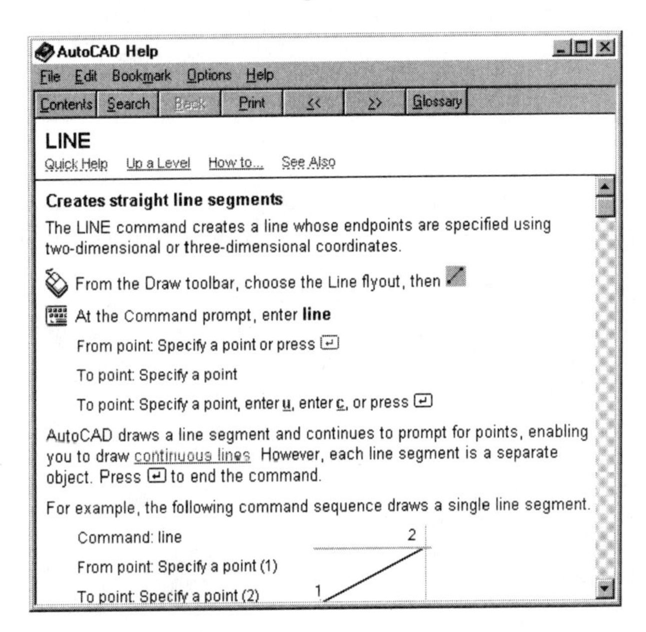

Figure 2

Notice that all instructions are highlighted with an arrow (>). This will make it easy for you to know exactly what you are expected to do. The comments that accompany the instructions are important for your understanding of what is happening and to help you avoid confusion.

TASK 2: Drawing Projects

There are three drawing projects at the end of each chapter. These are progressive, making use of previously learned commands as well as new ones. You will find the drawing itself on the right-hand page and drawing suggestions on the left-hand page.

Remember that any drawing may be executed in a number of ways. Our suggestions are not written in stone. Unlike the instructions in the introductory exercises, drawing instructions will not take you through the complete project. Besides the suggestions themselves, there is information on the drawing page that may assist you. In earlier chapters, this includes a list of commands you will need to use, a list of function keys and their functions, and, of course, the dimensions of the drawing itself. Later on you will see only the drawing and its dimensions.

TASK 3: Chapter 1

You are now ready to begin Chapter 1.

CHAPTER 1

COMMANDS

DRAW	EDIT	FILE	OPTIONS
LINE	U	NEW	UCSICON
	REDO	OPEN	
		QUIT	
		SAVE	
		SAVEAS	
VIEW			
REDRAW			

OVERVIEW

This chapter will introduce you to some of the tools you will use whenever you draw in AutoCAD. You will begin to find your way around the Release 13 for Windows menus and toolbars and you will learn to control basic elements of the drawing editor. You will produce drawings involving straight lines and learn to undo your last command with the U command. Your drawings will be saved, if you wish, using the SAVE or SAVEAS commands.

Look over the following tasks to get an idea of where we are going, and then begin Task 1.

TASKS

1. Begin a new drawing.
2. Explore the drawing editor.
3. Draw a line. Undo it, using the U command.
4. Review.

5. Draw a square. Use REDRAW to remove blips.
6. Save a drawing.
7. Do Drawing 1-1 ("Grate").
8. Do Drawing 1-2 ("Design").
9. Do Drawing 1-3 ("Shim").

TASK 1: Beginning a New Drawing

When you load Release 13 for Windows, you will find yourself in the drawing editor, which is where you do most of your work with AutoCAD. You can begin drawing immediately and name your drawing file later, or you can open a new or previously saved file. In this task, you will begin a new drawing and ensure that your drawing editor shows the "No Prototype" default settings we have used in preparing this chapter.

> From the Windows Program Manager, click on the AutoCAD Release 13 group.
> In the Release 13 group box, click on the AutoCAD Release 13 icon.
> Wait. . . .

You will see the Release 13 for Windows drawing editor, as shown in *Figure 1-1.* If other people have been using the computer you are using, it is quite possible that your screen will show other toolbar arrangements than the one shown here. We will take care of this shortly. If you are using an earlier version of AutoCAD, the screen

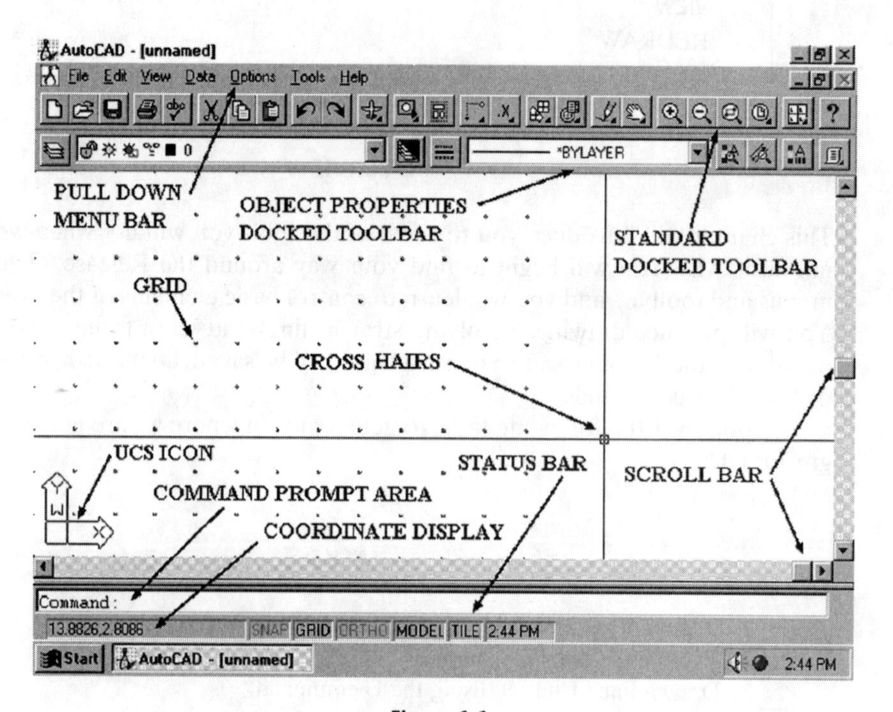

Figure 1-1

will certainly be different. In all cases, when you see the "Command:" prompt at the bottom of the screen you are ready to continue.

At this point, you could begin drawing. However, for our purposes it will be a good idea to use the NEW command to ensure that your drawing editor shows the same settings as are used in this chapter.

> Type "new" and press enter.

After you press enter, a dialogue box will appear, as shown in *Figure 1-2.* Dialogue boxes are used extensively in Release 13 and are discussed throughout this book. For now, simply notice the flashing vertical line in the rectangle at the bottom right. This indicates that AutoCAD is ready to accept typed input.

You will also see your cursor arrow somewhere on the screen. If you do not see it, move your pointing device to the middle of your drawing area or digitizer. When you see the arrow, you are ready to proceed.

If the box labeled "No Prototype" has an x in it, then you can skip the next step. If it is blank, then you will need to select it as follows:

> Move your pointing device until the arrow is in the box labeled "No Prototype" and press the pick button.

An X will appear in the box and "No Prototype" will be surrounded by dotted lines, indicating that this option has been selected.

> Move the arrow down to "OK" and click there. Typing <enter> will also work.

The dialogue box will disappear and your screen should resemble *Figure 1-1* again. You are now ready to proceed to Task 2.

TASK 2: Exploring the Drawing Editor

You are looking at the AutoCAD drawing editor. There are many ways that you can alter it to suit a particular drawing application. To begin with, there are a number of features that can be turned on and off using your mouse or the F-keys on your keyboard.

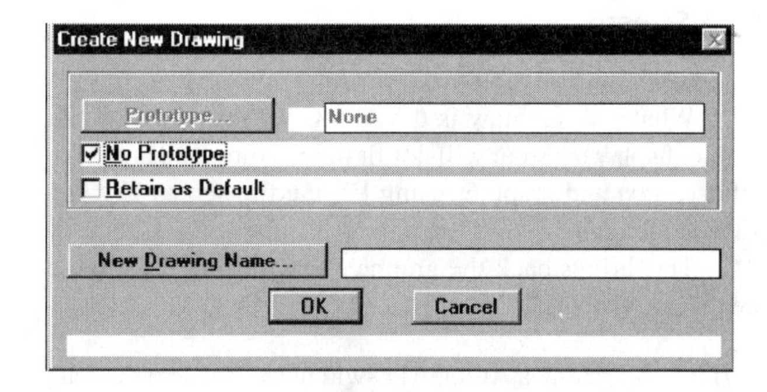

Figure 1-2

The Screen

At the top of the screen you will see the title bar, including the word "AutoCAD" followed by the name of the current drawing. Since your drawing is unnamed you will see "[unnamed]". This line also includes the standard Windows 95 close button on the left and the maximize and minimize buttons on the right. The close button will take you out of AutoCAD back to the program manager. The maximize and minimize buttons are used to switch between open applications and should be left alone while you are drawing in AutoCAD.

Below the title bar you will see the pull down menu bar including the titles for the File, Edit, View, Data, Options, Tools, and Help pull down menus. Pull down menus are discussed later in this task.

The next line down is probably the standard toolbar. This toolbar is one of many toolbars that can be displayed on the Release 13 Windows screen. The use of toolbars will also be discussed later in this chapter and in Chapter 3.

Typically, there will be one more line before you reach the drawing area. This is the Object Properties toolbar which displays the current layer and linetype. It also includes tools for changing other object properties. Layers and linetypes are discussed in Chapter 4. If you do not see this bar, don't worry because we are about to turn it off anyway.

Below these toolbars is the drawing area. You may also have one or more floating toolbars in the drawing area. These will be discussed later.

Continuing down the screen, below the drawing area and along the right side of the screen you will see scroll bars. These work like the scroll bars in any Windows application. Clicking on the arrows or clicking and dragging the square sliders will move your drawing to the left or right, up or down within the drawing area. You should have no immediate need for this function.

Beneath the scroll bar you will see the command prompt area. Typed commands are one of the basic ways of working in AutoCAD. The command line can be moved, resized, and reshaped like a toolbar. This is not recommended at this point.

At the bottom of the screen is the status bar, with the coordinate display on the left along with five mode indicators and the time display. The coordinate display and mode indicators are discussed shortly.

Switching Screens

> Press F2.

What you see now is the AutoCAD text window. AutoCAD uses this text window to display text that will not fit in the command area. You can switch back and forth between text and graphics using F2, the flip screen key.

> Press F2 again.

This brings back the graphics screen. Any text that does not fit in the command area is not visible.

NOTE: Sometimes AutoCAD switches to the text screen automatically when there is not enough room in the command area for prompts or messages. If this happens, use

F2 when you are ready to return to the graphics screen. Also, note that there is no ctrl key equivalent for F2.

Cross Hairs and Pickbox

You should see two lines at right angles horizontally and vertically, intersecting somewhere in the display area of your screen. If there are no cross hairs on your screen, move your pointing device until they appear. These are the cross hairs, or screen cursor, that tell you where your pointing device (puck, mouse, cursor, stylus, whatever) is located on your digitizer or mouse pad.

At the intersection of the cross hairs you will also see a small box. This is called the "pick box" and is used to select objects for editing. You will learn more about the pickbox later.

Move the pointer and see how the cross hairs move in coordination with your hand movements.

> Move the pointer so that the cross hairs move to the top of the screen.

When you leave the drawing area, your cross hairs will be left behind and you will see a pointing arrow. The arrow will be used as in other Windows applications to select tools and to pull down menus from the menu bar. Pull down menus are discussed at the beginning of Task 3.

NOTE: Here and throughout this book we show the Release 13 Windows 95 versions of AutoCAD screens in our illustrations. If you are working with another version, your screen will show significant variations.

> Move the cursor back into the drawing area and the selection arrow will disappear.

Toolbars and Pull Down Menus

There are fifty toolbars available in the standard AutoCAD Release 13 for Windows package. Toolbars can be created and modified. They can be moved, resized, and reshaped. They are a convenience, but they can also make your work and your drawing area overly cluttered. For our purposes, you will not need more than a few of the available toolbars.

Before going on, we will simplify your screen and ensure uniformity by turning off all toolbars and then turning on only the Standard and Draw toolbars.

> Move the selection arrow up to the word "Tools" in the pull down menu bar.

> Click the pick button (usually the button on the left) on your cursor.

This will open up the Tools menu as shown in *Figure 1-3.*

> Move the arrow down the menu to "Toolbars" and click.

The small black triangle to the right of "Toolbars" indicates that this selection calls up a submenu. Clicking on it will open up the Toolbars submenu as shown in the figure.

> Move the arrow all the way down the Toolbars submenu and click on "Close All".

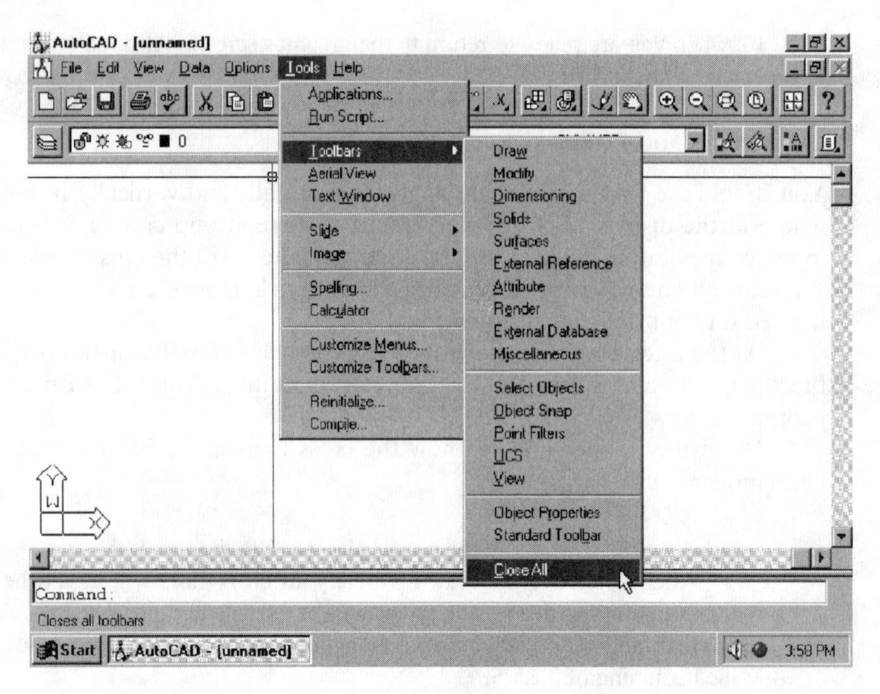

Figure 1-3

The menus will disappear along with any toolbars, and your screen will resemble *Figure 1-4.* This is the basic Release 13 for Windows drawing screen in its simplest form, with no toolbars or screen menu.

Now we will add the standard toolbar.

> Click on "Tools" again to open the Tools menu.

> Click on "Toolbars".

> Click on "Standard Toolbar" at the bottom of the menu, just above "Close All".

The Standard toolbar will be added to your screen as shown in *Figure 1-5.* Now add the Draw toolbar.

> Move the select arrow to the menu bar again and open the Tools pull down menu.

> Move down and open the Toolbars submenu.

> Click on "Draw".

The Draw toolbar will be added and your screen will resemble *Figure 1-6.*

The Coordinate Display

The coordinate display at the bottom left of the status line keeps track of coordinates as you move the pointer. The coordinate display is controlled using the F6 key, or by double clicking on the coordinate display itself.

Move the cursor around slowly and keep your eye on the pair of numbers at the bottom of the screen. They are probably moving very rapidly through four-place decimal numbers. When you stop moving, the numbers will be showing coordinates for the location of the pointer. These coordinates are standard ordered pairs in a coordi-

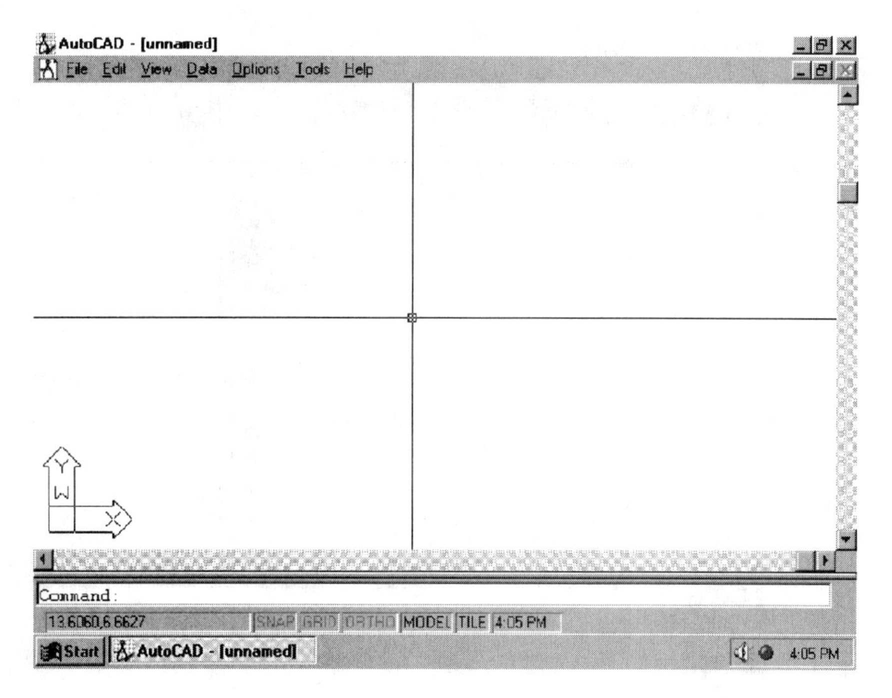

Figure 1-4

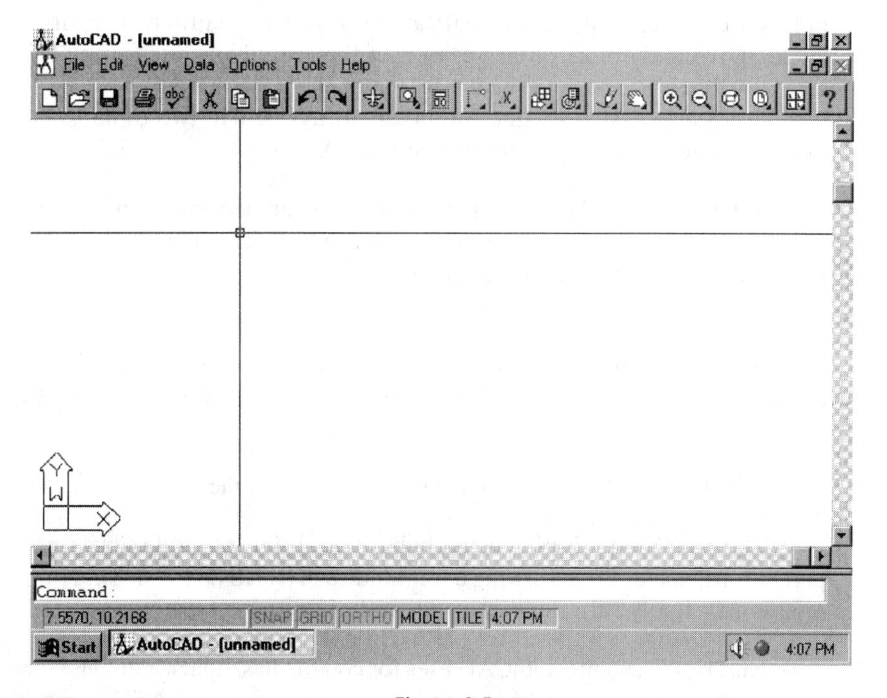

Figure 1-5

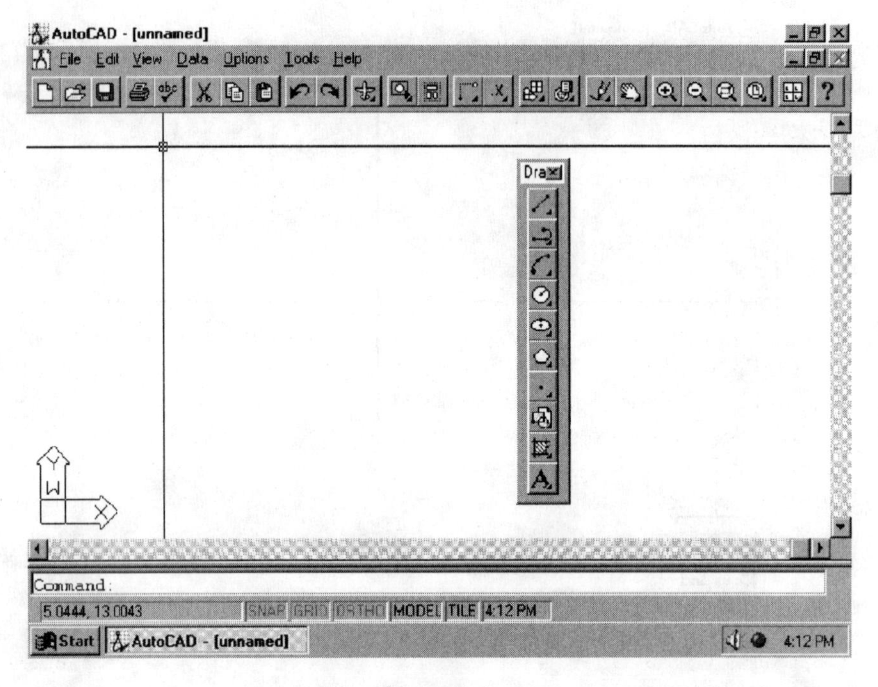

Figure 1-6

nate system originating from (0,0) at the lower left corner. The first value is the x value, showing the horizontal position of the cross hairs, measuring left to right. The second value is y, or the vertical position of the cross hairs, measured from bottom to top. Points also have a z value, but it will always be 0 in two-dimensional drawing and can be ignored until you begin to draw in 3D.

> Press F6 or double click on the coordinate numbers to turn the coordinate display off.

The numbers will freeze and the display will turn gray. You will also see "<Coords off>" on the command line.

> Move the cross hairs slowly.

Now when you move the cross hairs you will see that the coordinate display does not change. You will also see that it is grayed out, a standard Windows indication that this item is currently inactive or inaccessible.

> Press F6 or double click on the numbers to turn the coordinate display on again.

The coordinate display actually has two different modes, but this will not be apparent until you enter a drawing command such as LINE (Task 3). For now, F6 or your mouse will simply turn the display on and off.

NOTE: The units AutoCAD uses for coordinates, dimensions, and for measuring distances and angles can be changed at any time using the UNITS command (Chapter 2). For now we will accept the AutoCAD default values, including the four-place decimals.

In the next chapter we will be changing to two-place decimals. The F-keys are switches only; they cannot be used to change settings.

The Grid

> Press F7 or double click on the word GRID on the status bar to turn the grid on.

This will turn on the grid. When the grid is on, the word "GRID" will be black on the status bar.

The grid is simply a matrix of dots that helps you find your way around on the screen. It will not appear on your drawing when it is plotted, and it may be turned on and off at will. You may also change the spacing between dots, using the GRID command as we will be doing in Chapter 2.

The grid is presently set up to emulate the shape of an A-size (9×12) piece of paper. There are 10 grid points from bottom to top, numbered 0 to 9, and 13 points from left to right, numbered 0 to 12. The AutoCAD command that controls the outer size and shape of the grid is LIMITS, which will be discussed in Chapter 4. Until then we will continue to use the present format.

Also, you should be aware from the beginning that there is no need to scale AutoCAD drawings while you are working on them. That can be handled when you get ready to plot your drawing on paper. You will always draw at full scale, where one unit of length on the screen equals one unit of length in real space. This full-scale drawing space is called "model space". Notice the word "MODEL" on the status bar, indicating that you are in model space. The actual size of drawings printed out on paper is handled through "paper space". When you are in paper space you will see the word "PAPER" on the status bar. For now, all your work will be done in model space and you do not need to be concerned with paper space.

Snap

> Press F9 or double click on the word SNAP on the status bar.

The word should turn to black, indicating that snap mode is now on.

> Move the cursor slowly around the drawing area.

Notice how the cross hairs jump from point to point. If your grid is on, you will see that it is impossible to make the cross hairs touch a point that is not on the grid. Try it.

You will also see that the coordinate display shows only integer values, and that the word "SNAP" is displayed in black on the status bar.

> Press F9 or double click on "SNAP" again.

Snap should now be off and "SNAP" will be "grayed out" on the status bar again.

If you move the cursor in a circle now, you will see that the cross hairs move more smoothly without jumping. You will also observe that the coordinate display moves rapidly through a series of four-place decimal values.

F9 turns snap on and off. With snap off you can, theoretically, touch every point on the screen. With snap on, you can move only in predetermined increments. By default, the snap is set to a value of 1.0000. In the next chapter you will learn how to change this setting using the SNAP command. For now, we will leave it alone. A snap setting of 1 will be convenient for the drawings at the end of this chapter.

Using an appropriate snap increment is a tremendous timesaver. It allows for a degree of accuracy that is not possible otherwise. If all the dimensions in a drawing fall into one-inch increments, for example, there is no reason to deal with points that are not on a one-inch grid. You can find the points you want much more quickly and accurately if all those in between are temporarily eliminated. The snap setting will allow you to do that.

Ortho

F8 turns the ortho mode on and off. You will not observe the ortho mode in action until you have entered the LINE command, however. We will try it out at the end of Task 3.

The User Coordinate System Icon

At the lower left of the screen you will see the User Coordinate System (UCS) icon (see *Figure 1-1*). These two arrows clearly indicate the directions of the X and Y axes, which are currently aligned with the sides of your screen. In Chapter 9, when you begin to do 3D drawings, you will be defining your own coordinate systems that can be turned at any angle and originate at any point in space. At that time you will find that the icon is a very useful visual aid. However, it is hardly necessary in two-dimensional drawing and may be distracting. For this reason, you may want to turn it off now and keep it turned off until you actually need it.

> Type "ucsicon".

Notice that the typed letters are displayed on the command line to the right of the colon.

> Press "enter".

AutoCAD will show the following prompt on the command line:

ON/OFF/All/Noorigin/ORigin <ON>:

As you explore AutoCAD commands you will become familiar with many prompts like this one. It is simply a series of options separated by slashes (/). For now, we only need to know about "On" and "Off".

> Type "off" and press enter.

The UCS icon will disappear from your screen. Anytime you want to see it again, type in "ucsicon" and then type "on". Alternatively, you can click on "Options" on the menu bar, then on "UCS", and then on "Icon". This will turn the icon on if it is off and off if it is on.

TASK 3: Drawing a LINE

You can communicate drawing instructions to AutoCAD by typing or or by selecting items from a toolbar, a screen menu, or a pull down menu. Each method has its advantages and disadvantages, depending on the situation. Often, a combination of two or more methods is the most efficient way to carry out a complete command sequence. The instructions in this book are not always specific about which to use. All operators develop their own preferences.

Each method is described briefly. You do not have to try them all out at this time. Read them over to get a feel for the possibilities and then proceed to the LINE command. As a rule, we suggest learning the keyboard procedure first. It is the most basic, the most comprehensive, and changes the least from one release to the next. Do not limit yourself by typing everything. As soon as you know the keyboard sequence, try out the other methods to see how they vary and how you can use them to save time. Ultimately, you will want to type as little as possible and use the differences between the menu systems to your advantage.

The Keyboard and the Command Line

The keyboard is the most primitive and fundamental method of interacting with AutoCAD. Toolbars, Screen menus, and pull down menus all function by automating basic command sequences as they would be typed on the keyboard. It is, therefore, useful to be familiar with the keyboard procedures even if the other methods are sometimes faster.

As you type commands and responses to prompts, the characters you are typing will appear on the command line after the colon. Remember that you must press enter to complete your commands and responses. The command line can be moved and reshaped, or you can switch to the text screen using F2 when you want to see more lines including previously typed entries.

It is very useful to know that some of the most often used commands, such as LINE, ERASE, and CIRCLE, have "aliases." These one- or two-letter abbreviations are very handy. A few of the most commonly used aliases are shown in *Figure 1-7.*

COMMAND ALIAS CHART		
LETTER	+ ENTER	= COMMAND
A	↵	ARC
C	↵	CIRCLE
E	↵	ERASE
L	↵	LINE
M	↵	MOVE
P	↵	PAN
R	↵	REDRAW
Z	↵	ZOOM

Figure 1-7

Pull Down Menus

All of the menu systems and toolbars have the advantage that instead of typing a complete command, you can simply "point and shoot" to select an item. While the toolbars contain mostly drawing and editing commands, the pull down menus contain commands primarily used in the management and arrangement of your drawing. You will use many of its items and submenus frequently, and it is good to become familiar with what is there and what is not there.

To use the pull down menu, move the cross hairs up into the menu bar so that the selection arrow appears. Then move it to the menu heading you want. Select it with the pick button. A menu will appear. Run down the list of items to the one you want. Press the pick button again to select the item (see *Figure 1-8*). Items followed by a triangle have "cascading" submenus. Picking an item that is followed by an ellipsis (...) will call up a dialogue box (see *Figure 1-8*).

Dialogue boxes (see Task 6) are familiar features in many Windows and Macintosh programs. They require a combination of pointing and typing that is fairly intuitive. We will discuss many dialogue boxes in detail as we go along.

Toolbars

Toolbars are comprised of buttons with icons that give one-click access to many of AutoCAD's commands. Seventeen of the most commonly used toolbars can be opened directly from the Tools pull down menu in the same way that you opened the Standard

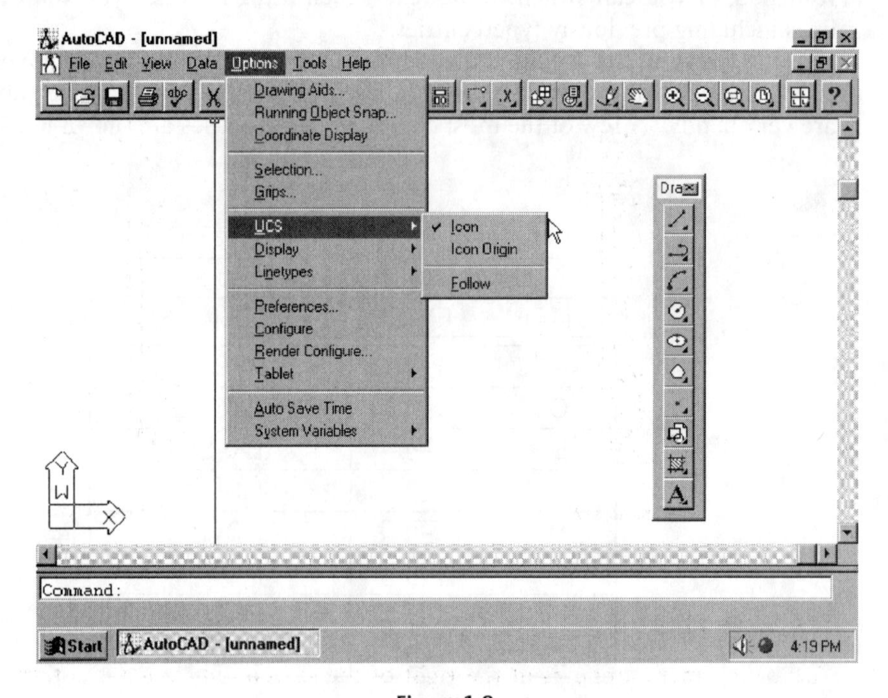

Figure 1-8

toolbar. Thirty-three others are available through the Customize Toolbars dialogue box, also on the Tools pull down menu.

Once opened, toolbars can "float" anywhere on the screen or can be "docked" along the edges of the drawing area. Toolbars can be a nuisance, since they cover portions of your drawing space. Do not use too many at once and remember that you can use the scroll bars to move your drawing right, left, up, and down behind the toolbars.

The icons used on toolbars are also a mixed blessing. One picture may be worth a thousand words, but with all the pictures you need to decipher you may find that a few words can be very handy as well. As in many other Windows applications, you can get a label for an icon simply by allowing the selection arrow to rest on the button for a moment without selecting it. These labels are called "Tooltips". Try this:

> Position the selection arrow on the top button of the Draw toolbar, as shown in *Figure 1-9,* but do not press the pick button.

You will see a yellow label that says "Line", as shown in the figure. This label identifies this button as the Line command tool. You will also notice that the status line now shows the phrase "Creates straight line segments". This is called a "helpstring".

The Line tool also has a small black triangle at the bottom right. This indicates that a "flyout" is available, giving you access to other related commands or variations on the Line command. If you press the pick button slowly and hold it down, the flyout will appear to the right. This flyout gives access to other variations of the LINE command, as shown in *Figure 1-10.*

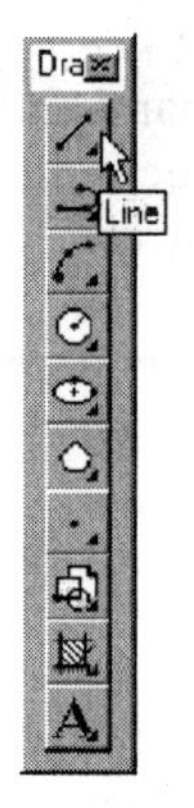

Figure 1-9

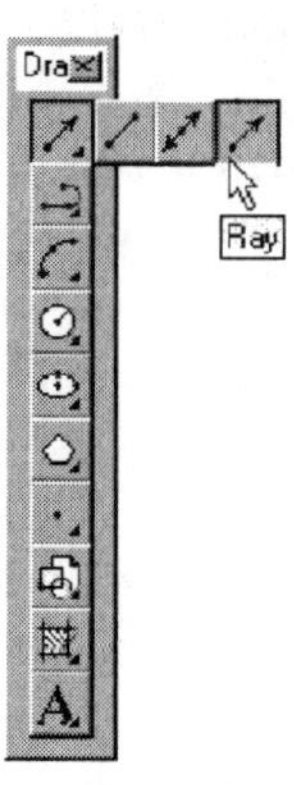

Figure 1-10

The Screen Menu

The screen menu appears at the right of the screen only when it is turned on in the Preferences dialogue box. We will not use it in this book.

The LINE Command

> Type "L" or select the Line tool from the Draw toolbar, as shown in *Figure 1-9* (remember to press enter if you are typing).

Look at the command area. You should see this, regardless of how you enter the command:

From point:

This is AutoCAD's way of asking for a start point.

Also, notice that the pickbox disappears from the cross hairs when you have entered a drawing command.

Most of the time when you are drawing you will want to point rather than type. In order to do this, you need to pay attention to the grid and the coordinate display.

> If snap is off, switch it on (F9 or double click on SNAP).

> Move the cursor until the display reads "1.0000,1.0000". Then press the pick button. If for any reason you cannot reach the point (1,1), use the scroll bars to move the grid in the drawing area.

Now AutoCAD will ask for a second point. You should see this in the command area:

To point:

Rubber Band

There are two new things to be aware of. One is the "rubber band" that extends from the start point to the cross hairs on the screen. If you move the cursor, you will see that this visual aid stretches, shrinks, or rotates like a tether keeping you connected to the start point. You will also notice that when the rubber band and the cross hairs overlap (i.e., when the rubber band is at 0, 90, 180, or 270 degrees) they both disappear in the area between the cross hairs and the start point, as illustrated in *Figure 1-11*. This may

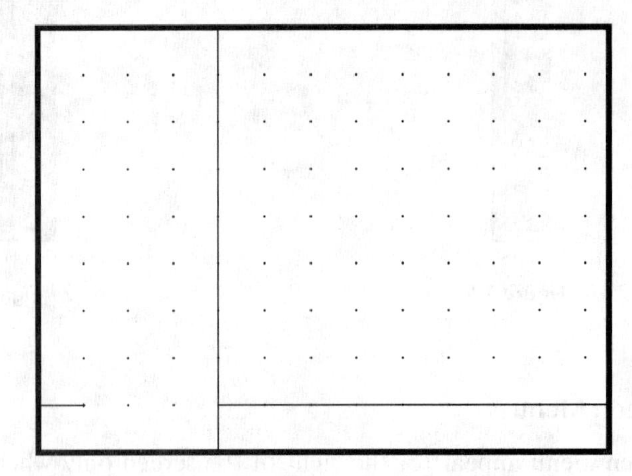

Figure 1-11

seem odd at first, but it is actually a great convenience. You will find many instances where you will need to know that the cross hairs and the rubber band are exactly lined up.

xy and Polar Coordinates

The other thing to watch is the coordinate display. If it is off (no change in coordinates when the cursor moves), press F6 to turn it on. Once it is on, it will show either xy coordinates or polar coordinates. xy coordinates are the familiar ordered pairs discussed previously. They are lengths measured from the origin (0,0) of the coordinate system at the lower left corner of the grid. If your display shows two four-digit numbers, then these are the xy coordinates.

If your display shows something like "4.2426<45", it is set on polar coordinates.

> Press F6 and move your cursor.
    Which type of coordinates is displayed?
> Press F6 and move your cursor again.
    Observe the coordinate display.
You will see that there are three coordinate display modes: off (no change), xy (x and y values separated by a comma), and polar (length<angle).
> Press F6 once or twice until it shows polar coordinates.

Polar coordinates are in a length, angle format, and are given relative to the starting point of your line. They look something like this: 4.0000<0 or 5.6569<45. The first number is the distance from the starting point of the line and the second is an angle of rotation, with 0 degrees being straight out to the right. In LINE, as well as most other draw commands, polar coordinates are very useful because they give you the length of the segment you are currently drawing.

> Press F6 to read the xy coordinates.
> Pick the point (8.0000,8.0000).
    Your screen should now resemble *Figure 1-12.* AutoCAD has drawn a line between (1,1) and (8,8) and is asking for another point.

To point:

This will allow you to stay in the LINE command to draw a whole series of connected lines if you wish. You can draw a single line from point to point, or a series of lines from point to point to point to point. In either case, you must tell AutoCAD when you are finished with a set of lines by pressing enter or the enter equivalent button on the cursor, or the space bar.

NOTE: When you are drawing a continuous series of lines, the polar coordinates on the display are given relative to the most recent point, not the original starting point.
> Press enter or the space bar to end the LINE command.

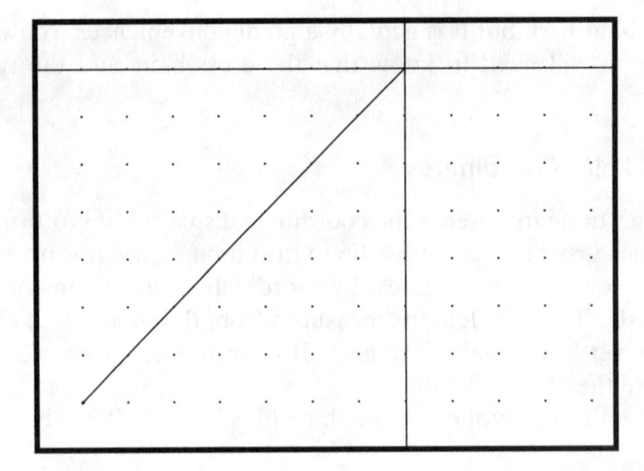

Figure 1-12

You should be back to the "Command:" prompt again, and the pickbox will have reappeared at the intersection of the cross hairs.

Space Bar and Enter Key

In most cases, AutoCAD allows you to use the space bar as a substitute for the enter or return key. This is a major convenience, since the space bar is easy to locate with one hand while the other hand is on the pointing device. For example, the LINE command can be entered without removing your right hand from your pointing device by typing "L" on the keyboard and then hitting the space bar, both with the left hand. The major exception to the use of the space bar as an enter key is when you are entering text in the TEXT command. Since a space may be part of a text string, the space bar must have its usual significance there.

NOTE: Hitting either the space bar or the enter key at the command prompt will repeat the last command entered, another major convenience.

Undoing a Line using U

> At the Command prompt, type "U" <enter> or select the Undo tool from the Standard toolbar, as shown in *Figure 1-13*.

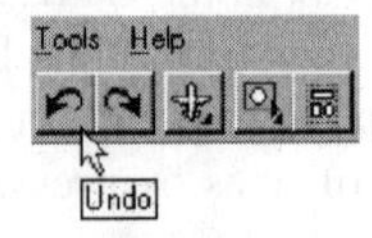

Figure 1-13

U undoes the last command, so if you have done anything else since drawing the line, you will need to type "U" <enter> more than once. In this way, you can walk backwards through your drawing session undoing your commands one by one.

> Type "REDO" <enter> or select the Redo tool (next to the Undo tool) immediately to bring the line back.

REDO only works immediately after U, and it only works once! That is, you can only REDO the last U and only if it was the last command executed.

Ortho

Before completing this section, we suggest that you try the ortho mode.

> Type "L" <enter> or select the Line tool to enter the LINE command.
> Pick a starting point. Any point near the center of the screen will do.
> Press F8 or double click on "ORTHO" and move the cursor in slow circles.

Notice how the rubber band jumps between horizontal and vertical without sweeping through any of the angles between. Ortho forces the pointing device to pick up points only along the horizontal and vertical quadrant lines from a given starting point. With ortho on, you can select points at 0, 90, 180, and 270 degrees of rotation from your starting point only (see *Figure 1-14*).

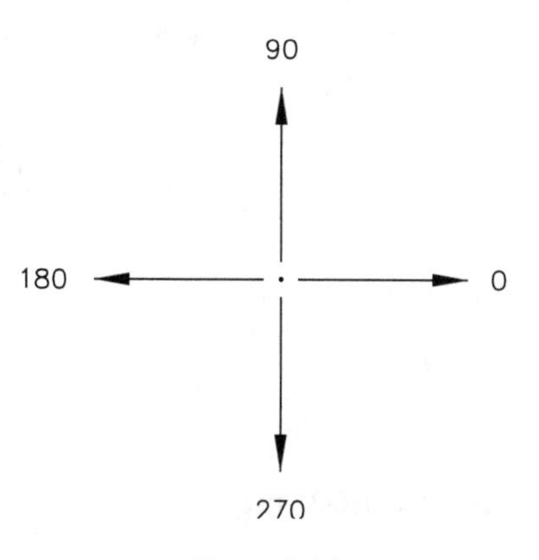

Figure 1-14

The advantages of ortho are similar to the advantages of snap mode, except that it limits angular rather than linear increments. It ensures that you will get precise and true right angles and perpendiculars easily when that is your intent. Ortho will become more important as drawings grow more complex. In this chapter it is hardly necessary, though it will be convenient in Drawings 1 and 3.

Escape

> While still in the LINE command, press the "Esc" (escape) key.

This will abort the LINE command and bring back the "Command": prompt. Esc is used to cancel a command that has been entered.

Object Selection Window

At some time during this chapter, it is likely that you will accidentally or intentionally pick a point on the screen without entering a command. If you then drag the cursor away from the point you just picked, a box will appear and stretch as you move. At the same time, AutoCAD will be prompting you for the "Other corner:" of the box but as soon as you pick it, the box will disappear. What is this? It's the object selection window that allows you to select objects for editing. We will begin using it in Chapter 2. You can cancel the prompt for the other corner by pressing Esc.

TASK 4: Review

Before going on to the drawings, quickly review the following items. If it all seems familiar, you should be ready for the next drawing task.

— F2 switches between text and graphics screens.
— F6 or double clicking on the coordinate display will turn coordinates on, off, or switch between polar and XY coordinates.
— F7 or double clicking on GRID turns the grid on and off.
— F8 or double clicking on ORTHO turns ortho on and off.
— F9 or double clicking on SNAP turns snap on and off.
— Commands may be entered by typing, or by selecting from a toolbar, a screen menu, or pull down menu.
— Points may be selected by pointing or by typing coordinates.
— Once inside the LINE command you can draw a single line or a whole series of connected lines.
— Press enter or the space bar to end working in the LINE command.
— U will undo your most recent command.
— REDO will redo a U.
— Esc will cancel a command.

TASK 5: Drawing and REDRAWing a Square

To practice what you have learned so far, reproduce the square in *Figure 1-15* on your screen. Then erase it using the U command as many times as necessary. The coordinates of the four corner points are (2,2), (7,2), (7,7), and (2,7).

REDRAW—Cleaning Up Your Act

You may have noticed that every time you select a point AutoCAD puts a "blip" on the screen in the form of a small cross. These are only temporary; they are not part of your drawing file database and will not appear on your drawing when it is plotted or printed. However, you will want to get rid of them from time to time to clean up the screen and avoid confusion.

> Type "r" <enter> or select the Redraw View tool from the Standard toolbar (illustrated in *Figure 1-16*).

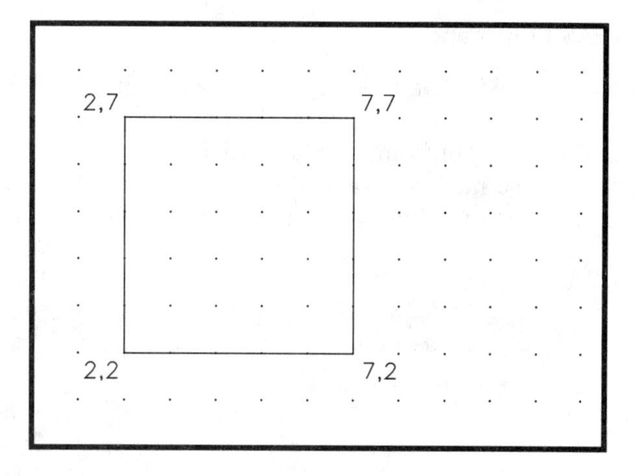

Figure 1-15

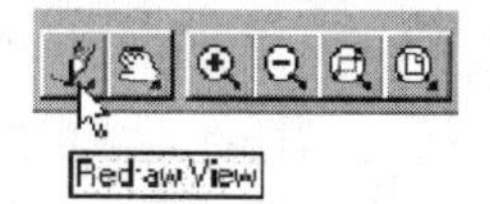

Figure 1-16

The display will be redrawn without the blips.

TASK 6: Saving Your Drawings

AutoCAD Release 13 for Windows has several commands that allow you to save drawings. Your choice of which to use will depend on whether you want to exit AutoCAD as you save or stay in the drawing editor, and whether you want to give the saved drawing a new name or keep it under the current drawing name. In all cases, a .dwg extension is added to file names to identify them as AutoCAD files. This is automatic when you name a file.

The SAVE Command

To save your drawing without leaving the editor, select "Save" from the File pull down menu or select the Save tool from the Standard toolbar (illustrated in *Figure 1-17*).

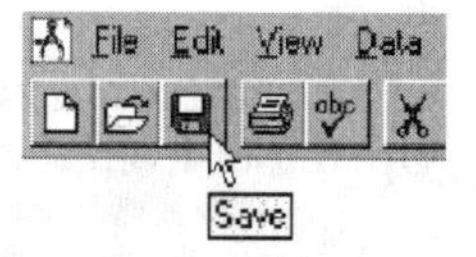

Figure 1-17

If the current drawing is already named, Release 13 will save it without intervening dialogue. If it is not named, it will open the SAVEAS dialogue box and allow you to give it a name to save it under.

The SAVEAS Command

To rename a drawing or to give a name to a previously unnamed drawing, type "Saveas", select "Saveas . . ." from the File pull down menu, or, in an unnamed drawing, select the Save tool from the Standard Toolbar.

Any of these methods will call up the Save Drawing As dialogue box (see *Figure 1-18*). The cursor will be blinking in the area labeled "Files:", waiting for you to

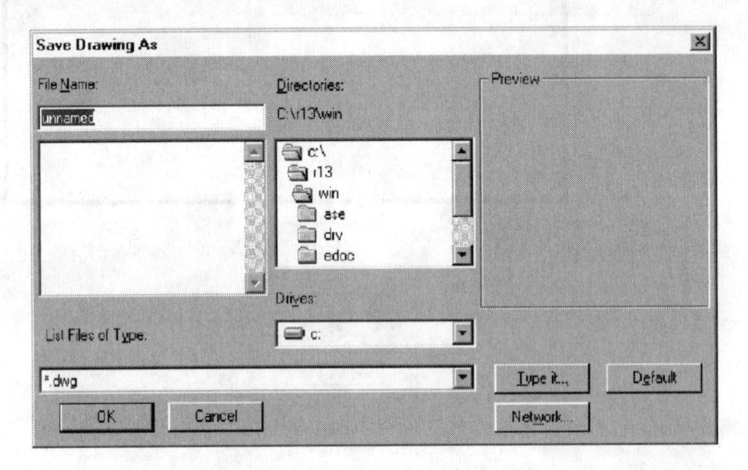

Figure 1-18

enter a file name. Include a drive designation (i.e., "A:1-1") if you are saving your work on a floppy disk. AutoCAD will add the .dwg extension automatically. SAVEAS will also allow you to save different versions of the same drawing under different names while continuing to edit.

The QUIT Command

To leave the drawing editor, type "quit" <enter> or select "Exit" from the File pull down menu.

A small dialogue box will appear. Pick "Save Changes . . . ," "Discard Changes," or "Cancel Command" as you wish.

OPENing Saved Drawings

To open a previously saved drawing, type "open", select "Open" from the File pull down menu, or select the Open tool from the Standard toolbar, shown in *Figure 1-19.*

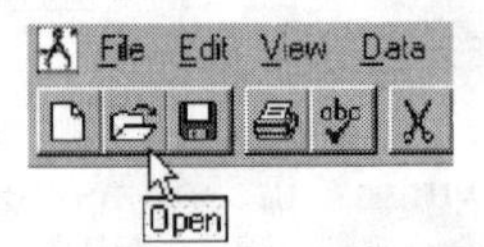

Figure 1-19

Once you have saved a drawing, you will need to use the OPEN command to return to it later. Entering the OPEN command by any method will bring up the Open Drawing dialogue box shown in *Figure 1-20.* You can select a file directory in the box

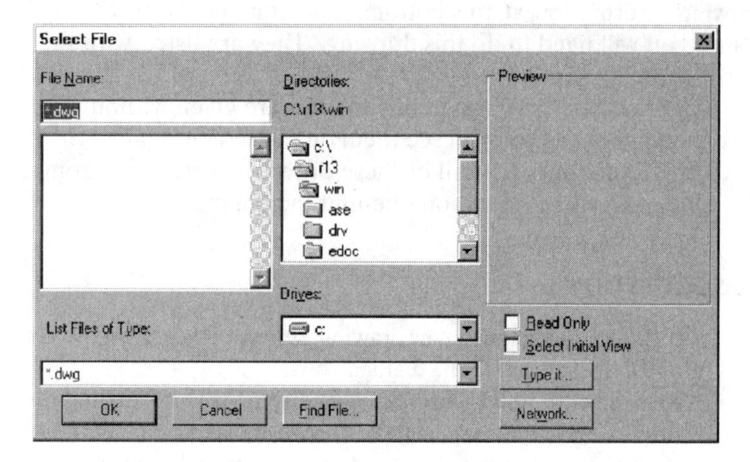

Figure 1-20

at the left and then a file from the box at the right. When you select a file, Release 13 will show a preview image of the selected drawing in the Preview image box at the right. This way you can make sure that you are opening the drawing you want.

TASKS 7, 8, and 9: Three Simple Drawings

You are now ready to complete this chapter by creating Drawings 1-1, 1-2, and 1-3. If you wish to save your drawings, use SAVE or SAVEAS. Whenever you wish to start a new drawing, type "new" and enter a file name and the No Prototype option as we did at the beginning of the chapter. Remember to use a drive designation if you are saving your work on a floppy disk.

DRAWING 1-1: GRATE

Before beginning, look over the drawing page. Notice the F-key reminders and other drawing information at the bottom. The commands on the right are the new commands you will need to do this drawing. They are listed with their screen menu headings in parentheses.

The first two drawings in this chapter are given without dimensions. Instead, we have drawn them as you will see them on the screen, against the background of a one-unit grid. Remember that all of these drawings were done using a one-unit snap, and that all points will be found on one-unit increments.

DRAWING SUGGESTIONS

> If you are beginning a new drawing, select "New . . ." on the "File" menu and then select "No Prototype" in the dialogue box.
> Remember to watch the coordinate display when searching for a point.
> Be sure that grid, snap, and the coordinate display are all turned on.
> Draw the outer rectangle first. It is six units wide and seven units high, and its lower left-hand corner is at the point (3.0000,1.0000). The three smaller rectangles inside are 4×1.

If You Make A Mistake—U

The U command works nicely within the LINE command to undo the last line you drew, or the last two or three if you have drawn a series.

> Type "U" <enter>. The last line you drew will be gone or replaced by the rubber band awaiting a new end point. If you want to go back more than one line, type "U" <enter> again, as many times as you need.
> If you have already left the LINE command, U will still work, but instead of undoing the last line, it will undo the last continuous series of lines. In Grate this could be the whole outside rectangle, for instance.
> Remember, if you have mistakenly undone something, you can get it back by typing "REDO" <enter>. You cannot perform other commands between U and REDO.

U is quick, easy to use, and efficient as long as you always spot your mistakes immediately after making them. Most of us, however, are more spontaneous in our blundering. We may make mistakes at any time and not notice them until the middle of next week. For us, AutoCAD provides more flexible editing tools, like ERASE, which is introduced in the next chapter.

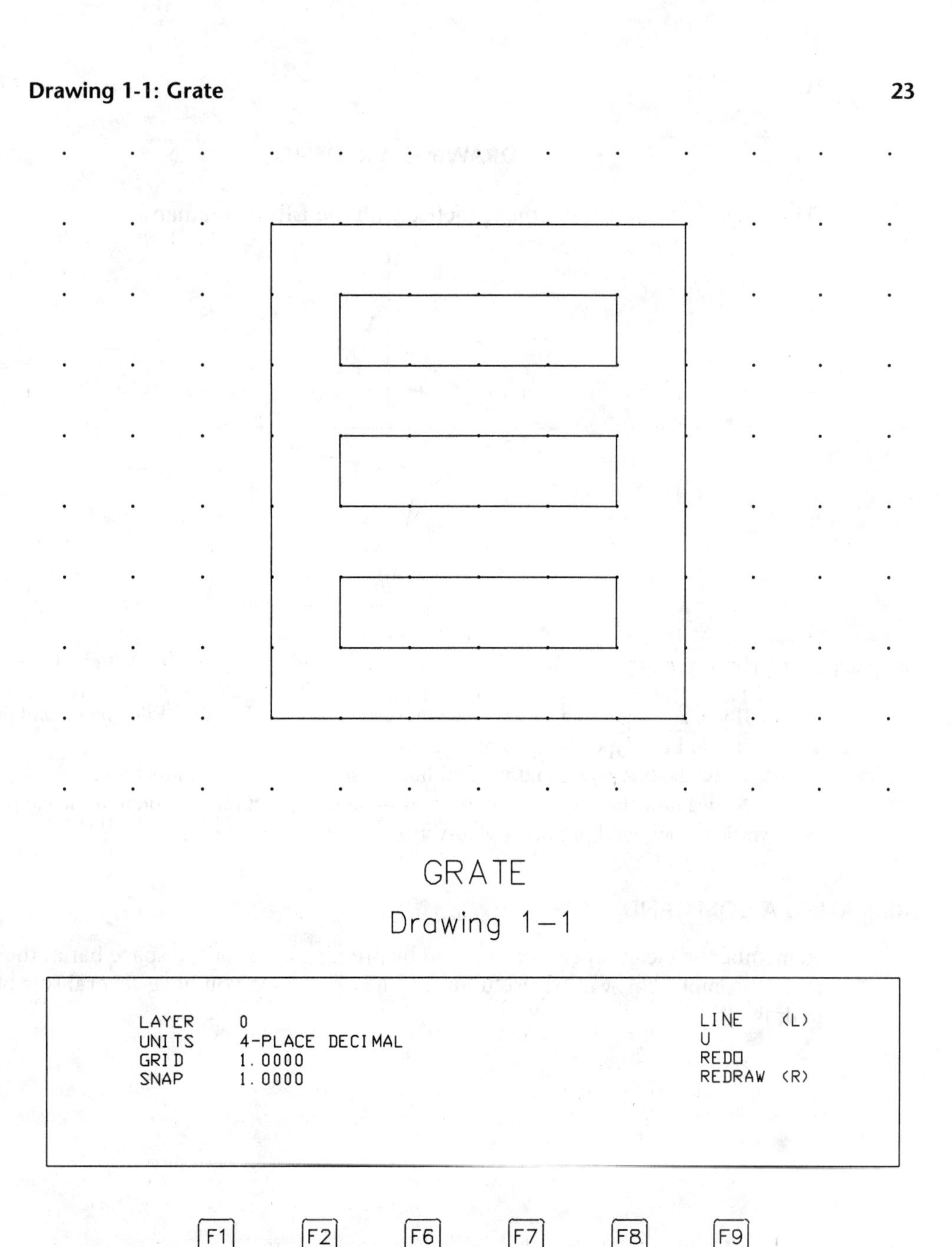
GRATE
Drawing 1–1
LAYER 0
UNITS 4-PLACE DECIMAL
GRID 1.0000
SNAP 1.0000
LINE (L)
U
REDO
REDRAW (R)
F1
HELP
F2
TEXT/GRAPHICS
SCREEN
F6
ABSOLUTE/OFF/POLAR
COORDS
F7
ON/OFF
GRID
F8
ON/OFF
ORTHO
F9
ON/OFF
SNAP

DRAWING 1-2: DESIGN

This design will give you further practice with the LINE command.

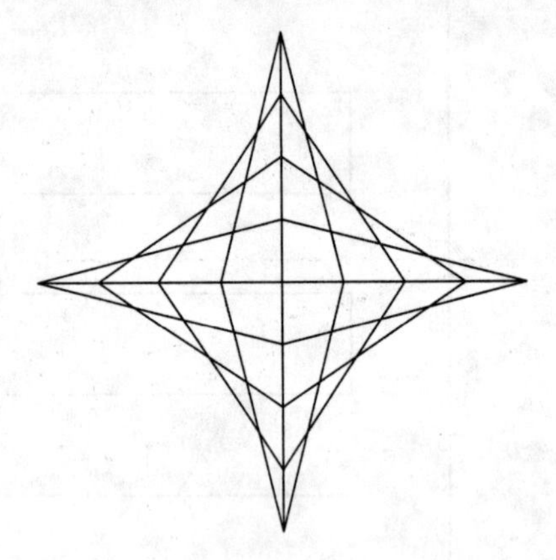

DRAWING SUGGESTIONS

> If you are beginning a new drawing, select "New . . ." on the "File" menu and then select "No Prototype" in the dialogue box.
> Draw the horizontal and vertical lines first. Each is eight units long.
> Notice how the rest of the lines work—outside point on horizontal to inside point on vertical, then working in, or vice-versa.

REPEATING A COMMAND

Remember, you can repeat a command by pressing enter or the space bar at the Command: prompt. This will be useful in this drawing, since you have several sets of lines to draw.

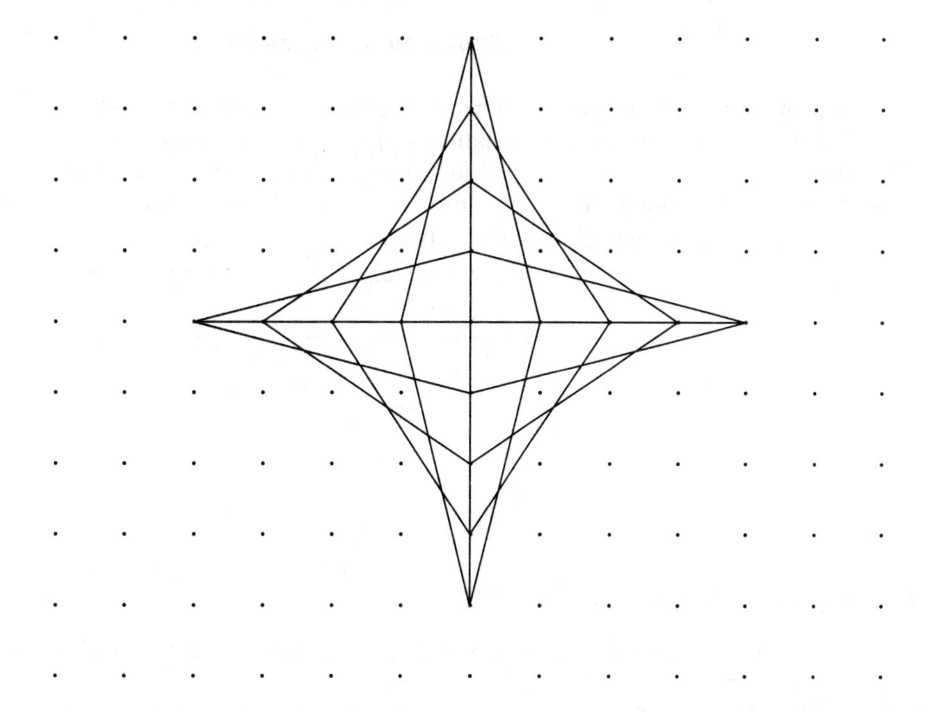

DESIGN # 1

Drawing 1-2

LAYER	0	LINE	(L)
UNITS	4-PLACE DECIMAL	U	
GRID	1.0000	REDO	
SNAP	1.0000	REDRAW	(R)

F1	F2	F6	F7	F8	F9
HELP	TEXT/GRAPHICS SCREEN	ABSOLUTE/OFF/POLAR COORDS	ON/OFF GRID	ON/OFF ORTHO	ON/OFF SNAP

DRAWING 1-3: SHIM

This drawing will give you further practice in using the LINE command. In addition, it will give you practice in translating dimensions into distances on the screen. Note that the dimensions are only included for your information; they are not part of the drawing at this point. Your drawing will appear like the reference drawing that follows. Dimensioning is discussed in Chapter 8.

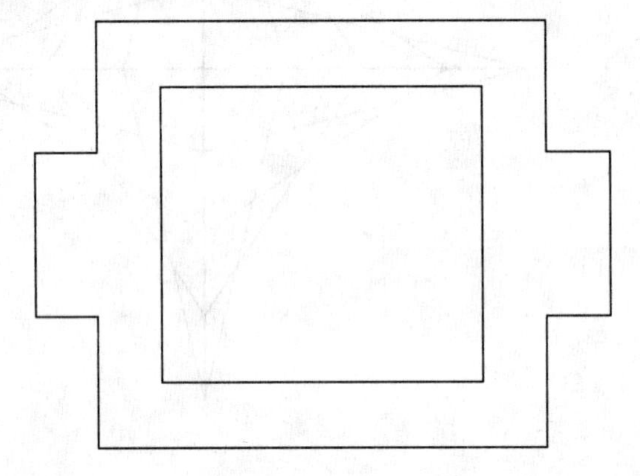

DRAWING SUGGESTIONS

> If you are beginning a new drawing, select "New . . ." on the "File" menu and then select "No Prototype" in the dialogue box.
> It is most important that you choose a starting point that will position the drawing so that it fits on your screen. If you begin with the bottom left-hand corner of the outside figure at the point (3,1), you should have no trouble.
> Read the dimensions carefully to see how the geometry of the drawing works. It is good practice to look over the dimensions before you begin drawing. Often, the dimension for a particular line may be located on another side of the figure or may have to be extrapolated from other dimensions. It is not uncommon to misread, misinterpret, or miscalculate a dimension, so take your time.

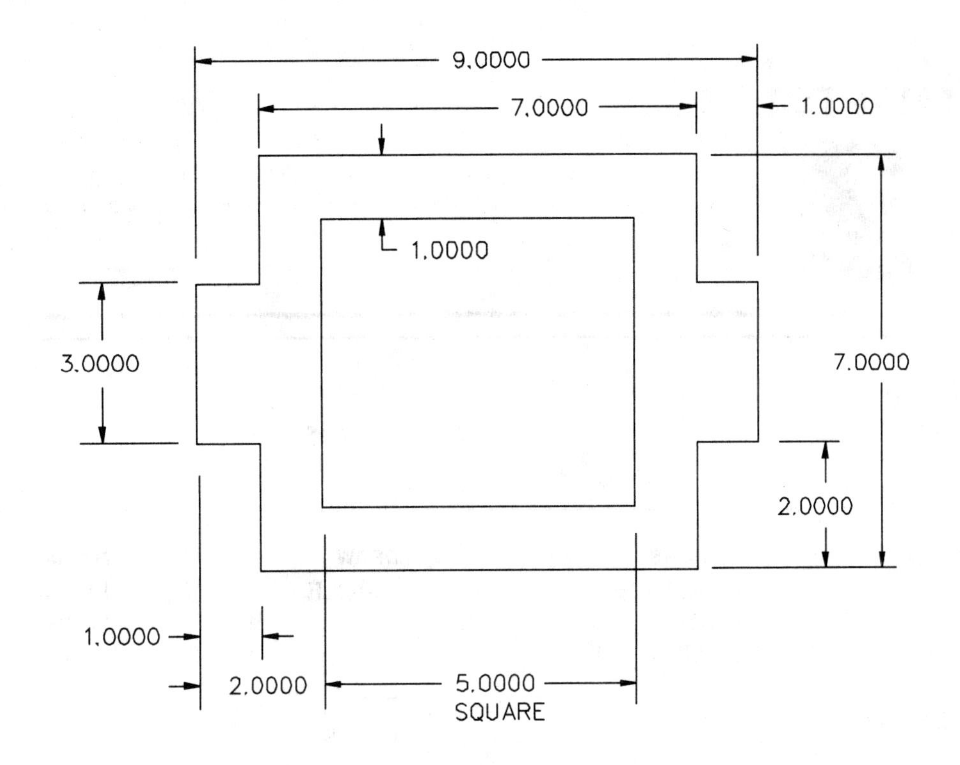

SHIM
Drawing 1-3

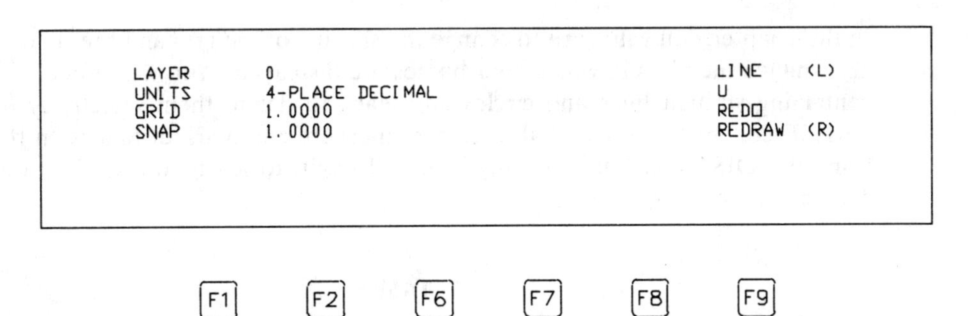

CHAPTER 2

COMMANDS

OPTIONS	DRAW	MODIFY
DDRMODES	CIRCLE	ERASE
GRID		OOPS
SNAP		
	DATA	
	UNITS	

OVERVIEW

In this chapter, you will learn to change the spacing of the grid and the snap. You will also change the units in which coordinates are displayed. You will produce drawings containing straight lines and circles and learn to delete them selectively using the ERASE command. You will also learn to measure and mark distances on the screen using the DIST command. Finally, you will begin to learn AutoCAD plotting and printing procedures.

TASKS

1. Change the snap spacing.
2. Change the grid spacing.
3. Change units.
4. Draw three concentric circles using the center point, radius method.
5. Draw three more concentric circles using the center point, diameter method.
6. ERASE the circles, using four selection methods.
7. Print or plot a drawing.

8. Do Drawing 2-1 ("Aperture Wheel").
9. Do Drawing 2-2 ("Roller").
10. Do Drawing 2-3 ("Switch Plate").

TASK 1: Changing the SNAP

When you begin a new drawing using no prototype (prototypes are discussed in Chapter 4), the grid and snap are set with a spacing of 1. Moreover, they are linked so that changing the snap will also change the grid to the same value. In Task 2 we will set the grid independently. For now, we will leave them linked.

In Chapter 1, all drawings were done without altering the grid and snap spacings from the prototype value of 1. Frequently you will want to change this, depending on your application. You may want a 10-foot snap for a building layout, or a 0.010-inch snap for a printed circuit diagram.

> To begin, type "new" or select "New" from the File pull down menu or the New tool from the Standard toolbar, as shown in *Figure 2-1*.

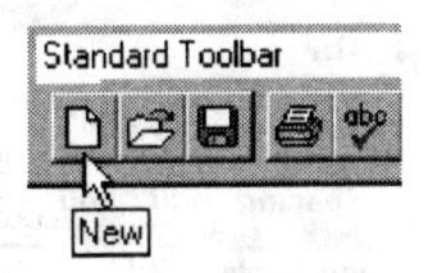

Figure 2-1

This will open the Create New Drawing dialogue box used in Chapter 1.

> Click on "No Prototype" to ensure that your drawing editor uses the default settings shown in this chapter.

> Using F7 and F9, be sure that grid and snap are both on.

> Type "snap" (we will no longer remind you to press enter after typing a command or response to a prompt). Do not use the pull down menu yet. We will get to those procedures momentarily.

The command area prompt will appear like this, with options separated by slashes (/):

Snap spacing or ON/OFF/Aspect/Rotate/Style <1.0000>:

You can ignore most of these options for now. The number <1.0000> shows the present setting. AutoCAD uses this format (<default>) in many command sequences to show you a present value or default setting. It usually comes at the end of a series of options. Pressing the enter key or space bar at this point will give you the default setting.

> In answer to the prompt, type ".5" and watch what happens (of course, you remembered to press enter).

Because the grid is set to change with the snap, you will see the grid redrawn to a .5 increment.

> Move the cursor around to observe the effects of the new snap setting.
> Try other snap settings. Try 2, .25, and .125. Remember that you can repeat the last command, SNAP, by pressing enter, the space bar, or the enter key on your pointing device.

How small a snap will AutoCAD and your video display accept? Notice that when you get too small (smaller than .08 on our screen), the grid becomes too dense to display but the snap can still be set smaller.

Using The Drawing Aids Dialogue Box

You can also change snap and grid settings using a dialogue box. The procedure is somewhat different, but the result is the same.

> Select "Options" from the pull down bar and then "Drawing Aids . . ." from the menu.
This method will call the Drawing Aids dialogue box shown in *Figure 2-2*.

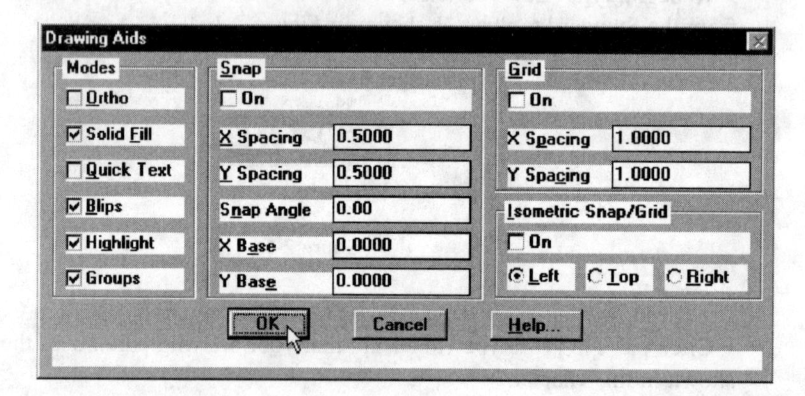

Figure 2-2

Look at the Drawing Aids dialogue box. It contains some typical features, including check boxes, edit boxes, and radio buttons.

Check Boxes

You can turn Ortho, Snap, and Grid on and off by moving the arrow inside the appropriate check box and pressing the pick button. A check box with a check is on, while an empty check box is off. Notice that there are five other check boxes in the area on the left under "Modes". We will only have use for the snap and grid settings at this point. If "Blips" is turned off, AutoCAD will not show blips on the screen as you draw. If "Highlight" is turned off, AutoCAD will not highlight selected entities. We will get to highlighting at the end of this chapter when we discuss the ERASE command.If "Groups" is not selected, then AutoCAD will not allow you to select grouped objects as a whole.

On the lower right, you will see a box for isometric snap and grid control. We will have no use for these in this book. Notice, however, the boxes in the lower line of the isometric snap/grid area labeled "Left", "Top", and "Right". These are examples of radio buttons, which are discussed in Task 3.

Also notice the "Help . . ." box at the bottom right. Again, any box with an ellipsis (. . .) calls up another dialogue box that will overlay the current one. Picking this box would activate the HELP command. Then, when you exit the HELP dialogue you would return to this Drawing Aids dialogue box.

Edit Boxes

The snap and grid settings are shown in edit boxes. Edit boxes contain text or numerical information that can be edited as you would in a text editor. You can highlight the entire box to replace the text, or point anywhere inside to do partial editing.

To change the grid or snap setting, use the following procedure:

1. Move the arrow into the box in the table where the change is to be made ("X Spacing" under "Snap" or "Grid").
2. Double click to highlight the entire number.
3. Type a new value and press enter. Notice that the Y spacing changes automatically when you change the X spacing (see the note that follows).
4. Click on the OK box at the bottom to confirm changes, or "Cancel" to cancel changes.

NOTE: The dialogue box has places to set both X and Y spacing. It is unlikely that you will want to have a grid or snap matrix with different horizontal and vertical increments, but the capacity is there if you do. Also notice that you can change the snap angle. Setting the snap angle to 45, for example, would turn your snap and grid at a 45 degree angle.

TASK 2: Changing the GRID

Whether you are typing or using one of your menus, the process for changing the grid setting is the same as changing the snap. In fact, the two are similar enough to cause confusion. The grid is only a visual reference. It has no effect on the selection of points. Snap is invisible, but it dramatically affects point selection. Grid and snap may or may not have the same setting.

> Using F7, be sure the grid is turned on.

> Type "grid" or use the Drawing Aids dialogue box, as discussed previously.

If you are typing, the prompt will appear like this, with options separated by slashes (/):

```
Grid spacing(X) or ON/OFF/Snap/Aspect <1.0000>:
```

> Type ".5".

> Try other grid settings. Try 2, .25, and .125. What happens when you try .0625?
> Now try setting snap to 1 and the grid spacing to .25. Notice how you cannot touch many of the dots. This is because the visible grid matrix is set to a smaller spacing than the invisible snap matrix.

In practice, this relationship is likely to be reversed. Since the grid is merely a visual aid, it will often be set "coarser" than the snap.

> Try setting the grid to .5 and the snap to .25.

With this type of arrangement you can still pick exact points easily, but the grid is not so dense as to be distracting.

If you wish to keep snap and grid the same, set the grid to "0". The grid will then change to match the snap and will continue to change any time you reset the snap. To free the grid, just give it its own value by again using the GRID command or the dialogue box.

TASK 3: Changing UNITS

> Type "Units" (do not use the pull down menu yet).

Presto! The text window appears and covers the graphics screen. We told you this would happen. The command area is too small to display the complete UNITS sequence, so it has been temporarily overlaid by the text screen. What function key will bring it back?

What you now see looks like this:

REPORT FORMATS:	**(EXAMPLES)**
1. Scientific	1.55E+01
2. Decimal	15.50
3. Engineering	1′−3.50″
4. Architectural	1′−3 1/2″
5. Fractional	15 1/2

With the exception of the Engineering and Architectural formats, these formats can be used with any basic unit of measurement. For example, decimal mode works for metric units as well as English units.

Enter choice, 1 to 5 <2>:

> Type "2" or simply press enter since the default system which we will use, is in decimal units.

Throughout most of this book we will stick to decimal units. Obviously, if you are designing a house you will want architectural units. If you are building a bridge, you may want engineering-style units. You might want scientific units if you are doing research.

Whatever your application, once you know how to change units, you can do so easily and at any time. However, as a drawing practice you will want to choose appropriate units when you first begin work on a new drawing. Not only will coordinates be displayed in the units you select, but later, when you use AutoCAD's dimensioning features (see Chapter 8), your drawing will be dimensioned in these units.

AutoCAD should now be showing the following prompt:

```
Number of digits to right of decimal point, (0 to 8) <4>:
```

We will use two-place decimals because they are practical and more common than any other choice.

> Type "2" in answer to the prompt for the number of decimal places you wish to use.

AutoCAD now gives you the opportunity to change the units in which angles are measured. In this book we will use all of the default settings for angle measure, since they are by far the most common. If your application requires something different, the UNITS command is the place to change it.

The default system is standard degrees without decimals, measured counterclockwise, with 0 being straight out to the right (3 o'clock), 90, straight up (12 o'clock), 180 to the left (9 o'clock), and 270 straight down (6 o'clock).

> Press enter four times, or until the "Command:" prompt reappears. Be careful not to press enter again, or the UNITS command sequence will be repeated.

Looking at the coordinate display, you should now see values with only two digits to the right of the decimal. This setting will be standard in this book.

We suggest that you experiment with other choices in order to get a feel for the options that are available to you. You should also try the Units Control dialogue box described next.

The Units Control Dialogue Box

Activate the Units Control dialogue box by picking "Units . . ." from the Data pull down menu. You will see the command DDUNITS entered on the command line, and the dialogue box shown in *Figure 2-3* will appear. This box shows all of the settings that are also available through the UNITS command sequence. There are two dialogue box features here that we have not discussed previously.

Radio Buttons

First are the radio buttons in the columns labeled "Units" and "Angles". Radio buttons are used with lists of settings that are mutually exclusive. You should see that the Decimal button is shaded in the format column. All other buttons in this column are not shaded. You can switch settings by simply picking another button, but you can have only one button on at a time. Radio buttons are used here because you can use only one format at a time.

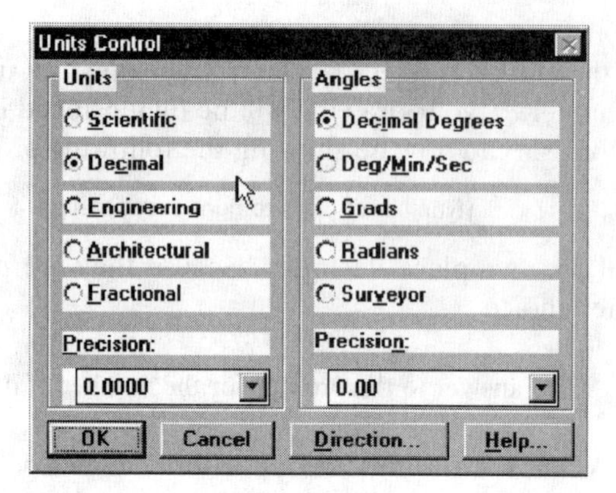

Figure 2-3

Pop Down Lists

The second new feature is the pop down lists at the bottom of the units and angles columns. You should try these out to see how they function. You can change the precision (number of place values shown) by picking the arrow at the right of the box and then picking a setting, such as "0.000" for three-place decimals, from the list that appears.

Experiment as much as you like with the dialogue box. None of your changes will be reflected in the drawing editor until you click on the OK button or press enter on the keyboard. When you are through, be sure to leave your units set for two-place decimals, your angle measure set for zero-place decimal degrees, and your angle zero direction set to East.

NOTE: All dialogue boxes can be moved on the screen. This is done by clicking in the gray title area at the top of the dialogue box, holding down the pick button, and dragging the box across the screen.

TASK 4: Drawing CIRCLES Giving Center Point and Radius

Circles can be drawn by giving AutoCAD three points on the circle's circumference, two points that determine a diameter, two tangents and a radius, a center point and a radius, or a center point and a diameter.

In this chapter, we will use the latter two options.

We will begin by drawing a circle with radius 3 and center at the point (6,5). Then, we will draw two smaller circles centered at the same point. Later, we will erase them using the ERASE command.

> Using what you have just learned, set grid and snap to .5 and units to two-place decimals.

> Type "c" or select the Circle tool from the Draw toolbar, illustrated in *Figure 2-4*. The prompt that follows will look like this:

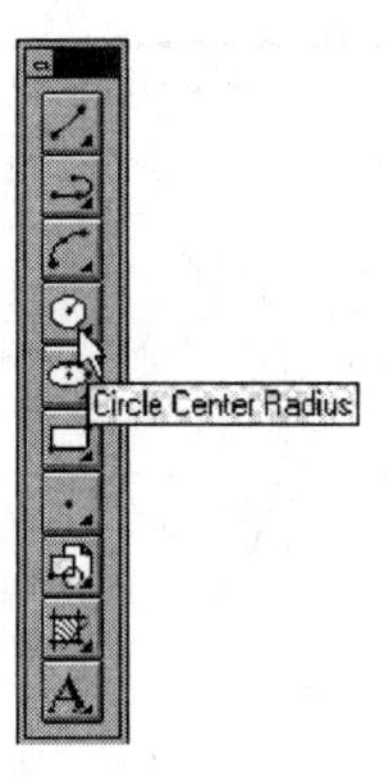

Figure 2-4

3P/2P/TTR/<Center point>:

> Type coordinates or point to the center point of the circle you want to draw. In our case, it will be the point (6,5). AutoCAD will assume that a radius or diameter will follow and will show the following prompt:

Diameter/<Radius>:

If we type or point to a value now, AutoCAD will take it as a radius, since that is the default.

> Type "3" or show by pointing that the circle has a radius of 3.

Notice how the rubber band works to drag out your circle as you move the cursor. Remember, if your coordinate display is not showing polar coordinates, press F6 once or twice until you see something like "3.00<0". When you are ready, press the pick button to show the radius end point.

You should now have your first circle complete. Next, draw two more circles using the same center point, radius method. They will be centered at (6,5) and have radii of 2.50 and 2.00. The results are illustrated in *Figure 2-5.*

Remember that you can repeat a command, in this case the CIRCLE command, by pressing enter or the space bar.

TASK 5: Drawing CIRCLES Giving Center Point and Diameter

We will draw three more circles centered on (6,5) having diameters of 2, 1.5, and 1. Drawing circles this way is almost the same as the radius method, except you will not use the default and you will see that the rubber band works differently.

> Press enter or the space bar to repeat the CIRCLE command. Type "c" or select the Circle tool from the Draw toolbar if you have done something else, such as a redraw, since drawing the first three circles. There is also a Circle Center Diameter tool on the Circle flyout, which we will explore momentarily.

> Indicate the center point (6,5) by typing coordinates or pointing.

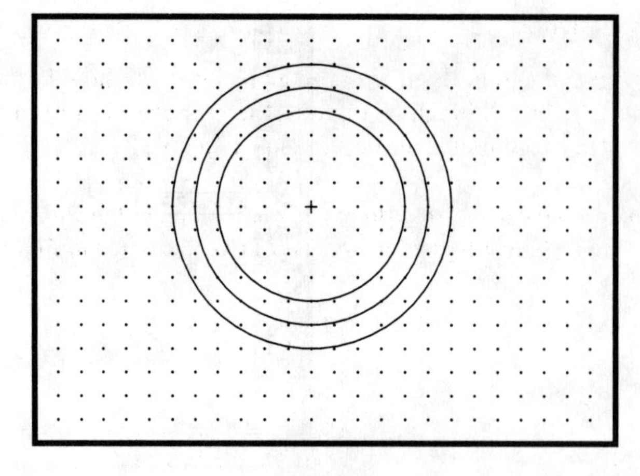

Figure 2-5

> Answer the prompt with a "d", for diameter.

Notice that the cross hairs are now outside the circle you are dragging on the screen (see *Figure 2-6*). This is because AutoCAD is looking for a diameter, but the last

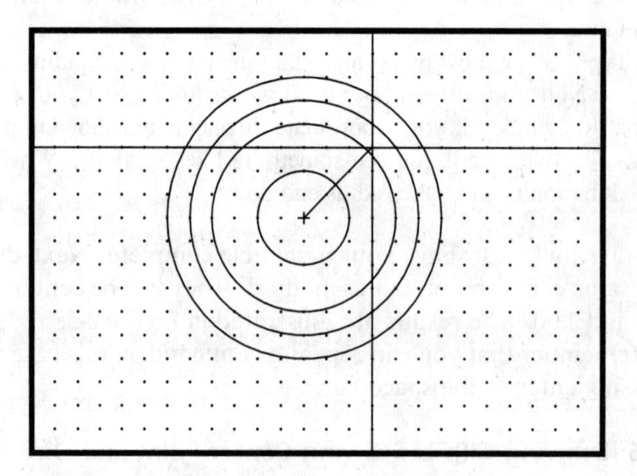

Figure 2-6

point you gave was a center point. So the diameter is being measured from the center point out, twice the radius. Move the cursor around, in and out from the center point, to get a feel for this.

> Point to a diameter of 2.00, or type "2".

You should now have four circles.

The Circle Flyout

We can use the Circle flyout menu bar to illustrate how flyouts work. As previously mentioned, the small black triangle in the corner of a tool button indicates that a flyout with other commands and options is available.

> Move the cursor arrow to the Circle tool, press the click button, and hold it down.
The Circle flyout bar will appear as illustrated in *Figure 2-7*.

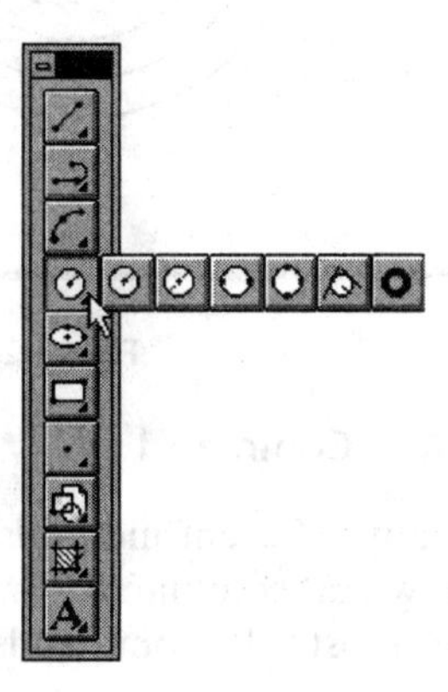

Figure 2-7

> Keep your finger on the button as you look at the flyout.
Notice that the first button is the same as the one on the main Draw toolbar. This is the case in all flyouts.
> Continue holding down the button and move the arrow to the right slowly.
Notice how each button is "held down" as the arrow passes over it.
> Move to the second button and let the arrow rest there.
You will see a yellow tooltip indicating that this is the Circle Center Diameter tool.
> Release the cursor button while the Circle Center Diameter tool is down.
This will initiate the CIRCLE command with the diameter option. In this case, "d" for diameter will be entered automatically at the second prompt. Briefly, this is because the menus and toolbars contain "macros" that automate what would otherwise be entered as keystrokes. If you watch the command area, you will see frequent examples of this.
Notice also that the Circle Center Diameter tool replaces the Circle Center Radius tool on the Draw toolbar as soon as you use it. It will stay there until you use the radius tool or another option again, or until you exit AutoCAD.
> Complete the command sequence to create a circle with a diameter of 1.5.

Using the diameter tool, draw one more circle with a diameter of 1.0. When you are done, your screen should look like *Figure 2-8*.

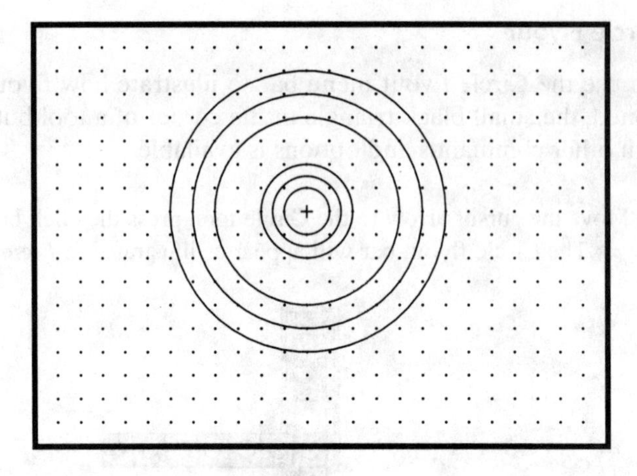

Figure 2-8

TASK 6: Using the ERASE Command

AutoCAD allows for many different methods of editing and even allows you to alter some of the basics of how edit commands work. Fundamentally, there are two different sequences for using most edit commands. These are called the Noun/Verb and Verb/Noun methods.

In earlier versions of AutoCAD, most editing was carried out in a verb/noun sequence. That is, you would enter a command, such as ERASE (the verb), then select objects (the nouns), and finally press enter to carry out the command. This method is still perfectly reasonable and effective, but AutoCAD now allows you to reverse the verb/noun sequence. You can use either method as long as the "Noun/Verb" selection is enabled in your drawing.

In this task, we will explore the traditional verb/noun sequence and then introduce the noun/verb or "pick first" method along with some of the many methods for selecting objects.

Verb/Noun Editing

> To begin this task you should have the six circles on your screen, as shown previously in *Figure 2-8*.

We will use verb/noun editing to erase the two outer circles. If you wish to be able to use the Erase tool, you will have to first open the Modify toolbar as follows:

1. Open the Tools pull down menu.
2. Open the Toolbars submenu.
3. Click on Modify.

> Type "e" or select the Erase tool from the Modify toolbar, as shown in *Figure 2-9*.

The cross hairs will disappear, but the pickbox will still be on the screen and will move when you move your cursor.

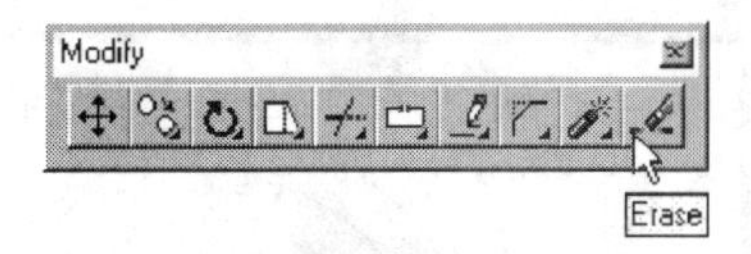

Figure 2-9

Also notice the command area. It should be showing this:

Select objects:

This is a very common prompt. You will find it in all edit commands and many other commands as well.
> Move your cursor so that the outer circle crosses the pickbox.

NOTE: In many situations, you may find it convenient or necessary to turn snap off (F9) while selecting objects since this gives you more freedom of motion.

> Press the pick button.

The circle will be highlighted (dotted). This is how AutoCAD indicates that an object has been selected for editing. It is not yet erased, however. You can go on and add more objects to the selection set and they, too, will become dotted.
> Use the box to pick the second circle. It too should now be dotted.
> Press enter, the space bar, or the enter equivalent button on your pointing device to carry out the command.

This is typical of the verb/noun sequence in most edit commands. Once a command has been entered and a selection set defined, a press of the enter key is required to complete the command. At this point, the two outer circles should be erased.

Noun/Verb Editing

Now let's try the noun/verb sequence.
> Type "u" to undo the ERASE and bring back the circles.
> To ensure that noun/verb editing is enabled in your drawing, pick "Options" from the pull down menu and then "Selection . . .".

This will open the Object Selection Settings dialogue box.
> If the Noun/Verb Selection check box is not checked, click in the box to check it.
> Click on "OK" to exit the dialogue.

You are now ready to use pick first editing.
> Use the pickbox to select the outer circle.

The circle will be highlighted, and your screen should now resemble *Figure 2-10.* Those little blue boxes are called "grips". They are part of AutoCAD's autoediting system, which we will begin exploring in Chapter 3. For now, you can ignore them.
> Pick the second circle in the same fashion.

The second circle will also become dotted and more grips will appear.

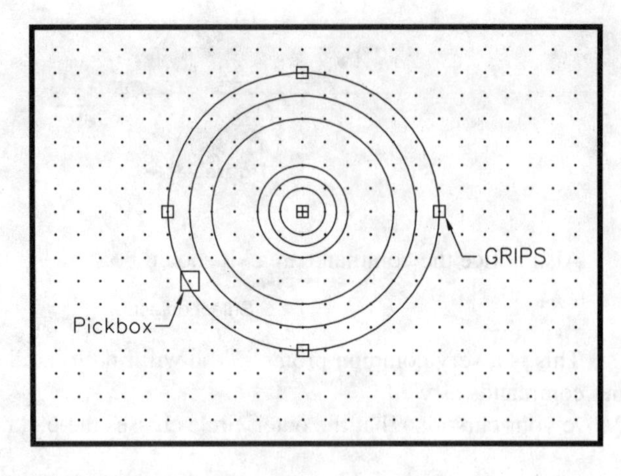

Figure 2-10

> Type "e" or select the Erase tool.

Your two outer circles will disappear as soon as you press enter or pick the tool.

The two outer circles should now be gone. As you can see, there is not a lot of difference between the two sequences. One difference that is not immediately apparent is that there are a number of selection methods available in the older verb/noun system that cannot be activated when you pick objects first. We will get to other object select methods momentarily, but first try out the OOPS command.

OOPS!

> Type or select "Oops" and watch the screen.

If you have made a mistake in your erasure, you can get your selection set back by typing (or selecting) "Oops". OOPS is to ERASE as REDO is to UNDO. You can use OOPS to undo an ERASE command, as long as you have not done another ERASE in the meantime. In other words, AutoCAD only saves your most recent ERASE selection set.

You can also use U to undo an ERASE, but notice the difference: U simply undoes the last command, whatever it might be; OOPS works specifically with ERASE to recall the last set of erased objects. If you have drawn other objects in the meantime, you can still use OOPS to recall a previously erased set. If you tried to use U, you would have to backtrack, undoing any newly drawn objects along the way.

Other Object Selection Methods

You can select individual entities on the screen by pointing to them one by one, as we have done, but in complex drawings this will often be inefficient. AutoCAD offers a variety of other methods, all of which have application in specific drawing circumstances. In this exercise we will select circles by the "windowing" and "crossing" methods, and by indicating "last" or "L", meaning the last entity drawn.

We suggest that you study *Figure 2-11* to learn about other methods. The number of selection options available may seem a bit overwhelming at first, but the time you spend learning them will be well spent. These same options will appear in numerous AutoCAD editing commands (MOVE, COPY, ARRAY, ROTATE, MIRROR) and should become part of your CAD vocabulary.

OBJECT SELECTION METHOD	DESCRIPTION	ITEMS SELECTED
(W) WINDOW	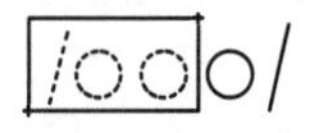	THE ENTITIES WITHIN THE BOX
(C) CROSSING	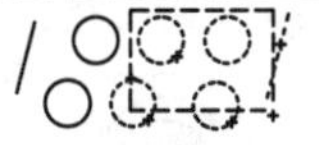	THE ENTITIES CROSSED BY OR WITHIN THE BOX
(P) PREVIOUS		THE ENTITIES THAT WERE PREVIOUSLY PICKED
(L) LAST		THE ENTITY THAT WAS DRAWN LAST
(R) REMOVE	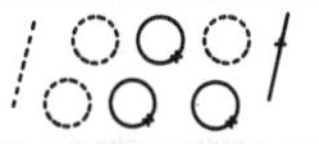	REMOVES ENTITIES FROM THE ITEMS SELECTED SO THEY WILL NOT BE PART OF THE SELECTED GROUP
(A) ADD	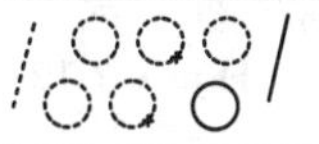	ADDS ENTITIES THAT WERE REMOVED AND ALLOWS FOR MORE SELECTION AFTER THE USE OF REMOVE
ALL	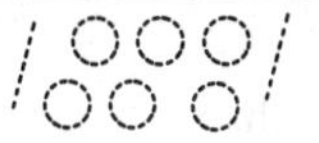	ALL ENTITIES CURRENTLY VISIBLE ON THE DRAWING
(F) FENCE		THE ENTITIES CROSSED BY THE FENCE
(WP) WPOLYGON	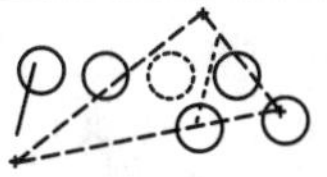	THE ENTITIES WITHIN THE THE POLYGON
(CP) CPOLYGON	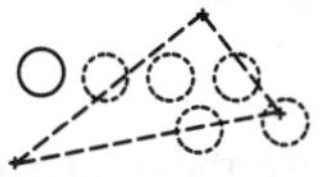	THE ENTITIES CROSSED BY OR WITHIN THE PLOYGON

Figure 2-11

Selection by Window

In Releases 12 and 13, window and crossing selections, like pointing to individual objects, can be initiated without entering a command. In other words, they are available for noun/verb selection. Also, whether you select objects first or enter a command first, you can force a window or crossing selection simply by picking points on the screen that are not on objects. AutoCAD will assume you want to select objects and will ask for a second point.

Let's try it. We will show AutoCAD that we want to erase all of the inner circles by throwing a temporary selection window around them. The window will be defined by two points moving left to right that serve as opposite corners of a rectangle. Only entities that lie completely within the window will be selected (see *Figure 2-12*).

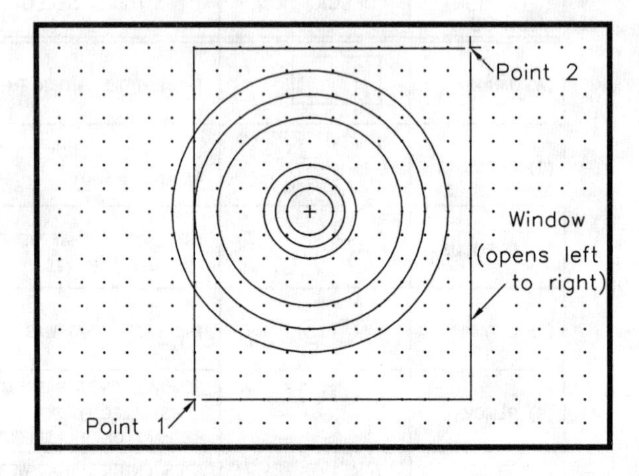

Figure 2-12

> Pick point 1 at the lower left of the screen, as shown. Any point in the neighborhood of (3.5,1) will do.

AutoCAD will prompt for another corner:

Other corner:

> Pick point 2 at the upper right of the screen, as shown. Any point in the neighborhood of (9.5,8.5) will do. To see the effect of the window, be sure that it crosses the outside circle as in *Figure 2-12.*

> Type "e" or select the Erase tool.

The inner circles should now be erased.

> Type or select "Oops" to retrieve the circles once more. Since ERASE was the last command, typing "U" will work equally well.

Selection by Crossing Window

Crossing is an alternative to windowing that is useful in many cases where a standard window selection could not be performed. The selection procedure is the same, but a crossing box opens to the left instead of to the right and all objects that cross the box will be chosen, not just those that lie completely inside the box.

We will use a crossing box to select the inside circles.

> Pick point 1 close to (8.0,3.0) as in *Figure 2-13*.
AutoCAD prompts:

Other corner:

> Pick a point near (4.0,7.0). This point selection must be done carefully in order to demonstrate a crossing selection. Notice that the crossing box is shown with dotted lines, whereas the window box was shown with solid lines.

Also, notice how the circles are selected: those that cross and those that are completely contained within the box, but not those that lie outside.

At this point we could enter the ERASE command to erase the circles, but instead we will demonstrate how to use the Esc key to cancel a selection set.

> Press the Esc key on your keyboard. This will cancel the selection set. The circles will no longer be highlighted, but you will see that the grips are still visible. To get rid of the grips, you will need to cancel again.

> Press Esc again.

The grips should now be gone as well.

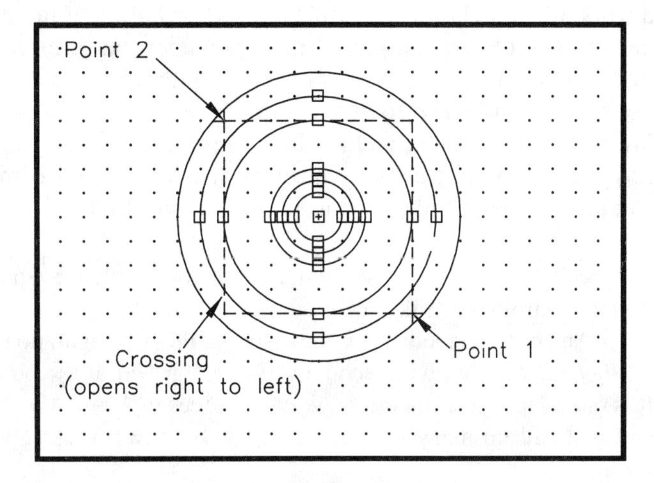

Figure 2-13

Selecting the "Last" Entity

AutoCAD remembers the order in which new objects have been drawn during the course of a single drawing session. As long as you do not leave the drawing editor, you can select the last drawn entity using the "last" option. If you leave the drawing editor and return later, this information will no longer be available.

> Type "e" or select the Erase tool.

Notice that there is no way to specify "last" before you enter a command. This option is only available as part of a command procedure. In other words, it only works in a verb/noun sequence.

> Type "L".

The inner circle should be highlighted.

> Press enter to carry out the command.

The inner circle should be erased.

There are several more object selection methods, all of which may be useful to you in a given situation. Some of the most useful options are described and illustrated in *Figure 2-11,* which you should study before moving on.

TASK 7: Plotting or Printing a Drawing

AutoCAD's printing and plotting capabilities are extensive and complex. In this book we will introduce you to them a little bit at a time. These presentations are intended to get your work out on paper efficiently.

For starters, we will show you how to do a very simple plot procedure. This procedure assumes that you do not have to change devices or configuration details. It should work adequately for the drawings in this chapter. One of the difficulties is that different types of plotters and printers work somewhat differently. We will try to present procedures that will work on whatever equipment you are using, assuming that it is appropriately configured to begin with.

We suggest that you open the Plot Configuration dialogue box and work through this section now without actually plotting anything. Then come back to it after you have completed one of the drawings in this chapter that you want to print out.

> Type "plot" or Ctrl-p, select the Print tool from the Standard toolbar, or select "Print..." from the "File" menu.

Any of these methods will call up the Plot Configuration dialogue box illustrated in *Figure 2-14.* You will become very familiar with it as you work through this book. It is one of the most important working spaces in AutoCAD. It contains many options and will call up many subdialogues, but for now we are going to look at only two settings.

Look at the box on the lower left labeled "Additional Parameters". The line of radio buttons on the left allows you to tell AutoCAD what part of your drawing you want

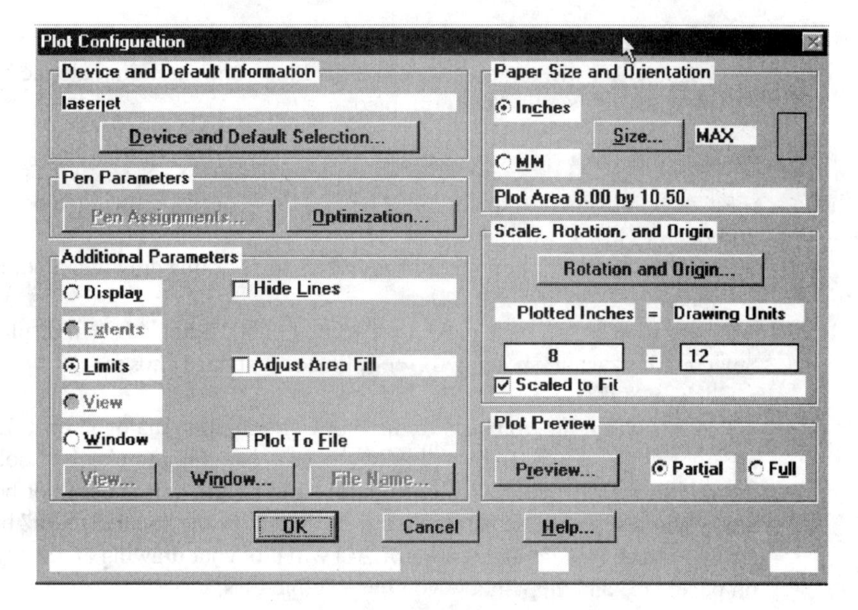

Figure 2-14

to plot. For our purposes we will use "Window". For this exercise you will define a window around an object you have drawn. Windowing allows you to plot any portion of a drawing simply by defining the window. AutoCAD will base the size and placement of the printed drawing on the window you have defined on the screen. For now, any object will do.

> Click on "Window . . ." at the bottom middle of the Additional Parameters box.

This will call up the Window Selection dialogue box illustrated in *Figure 2-15*. It contains the X and Y coordinates of the two corners of your window selection. To ensure that you get the window you want, you need to pick the two points or type in the coordinates if you know them. We will pick points, using Drawing 2-1 to illustrate.

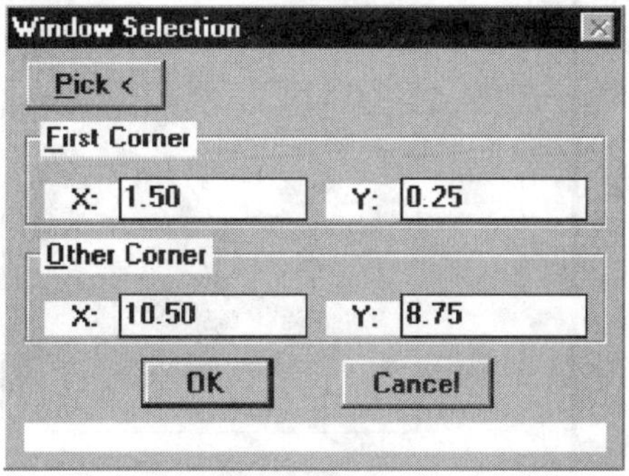

Window Selection
Pick <
First Corner
X: 1.50 Y: 0.25
Other Corner
X: 10.50 Y: 8.75
OK Cancel

Figure 2-15

> Click on "Pick <" at the top left of the dialogue box.

AutoCAD will temporarily close the Plot Configuration dialogue box so that you can pick points. You will be prompted:

First corner:

> Pick a point similar to point 1 in *Figure 2-16.* For this drawing, a point in the neighborhood of (1.5,.25) will do.
> Pick a point similar to point 2 in *Figure 2-16.* For this drawing, a point in the neighborhood of (10.5,8.75) will do.

As soon as you have picked the second point, AutoCAD will display the Window Selection box again, showing the coordinates you have chosen.
> Click on "OK".

This will bring you back to the Plot Configuration dialogue box. Essentially you are done at this point. Before plotting, there is one thing to check. Look at the box at the middle right labeled "Scale, Rotation, and Origin". It should not be necessary to worry about any of these parameters at this point. If the "Scaled to Fit" box is checked, as it should be by default, then AutoCAD will plot your drawing at maximum size based on paper size and the window you have defined.
> If "Scaled to Fit" is not checked, click in the box to check it.
> If you are not actually going to print a drawing at this point, click on Cancel to close the Plot Configuration dialogue box.

At this point the exercise is complete. You should come back to it when you have a drawing ready to plot.
> If you are ready to proceed with a plot, look at your printer or plotter. Make sure that it is on, online, and that the paper is ready to go.
> Click on "OK".

This sends the drawing information out to be printed. You can sit back and watch the plotting device at work. If you need to cancel for any reason, click the Cancel button.

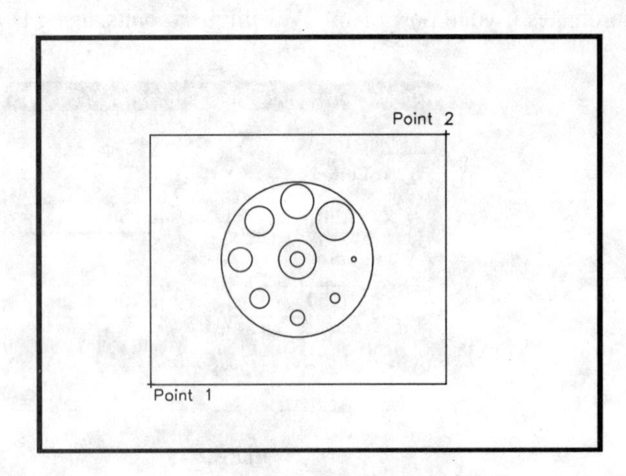

Figure 2-16

For now, that's all there is to it. If your plot does not look perfect (if it is not centered, for example), do not worry, you will learn all you need to know about plotting later on.

TASKS 8, 9, and 10

You are now ready to complete Drawings 2-1 through 2-3. Remember to set grid, snap, and units before you begin each drawing. Use either ERASE or U if you make a mistake, depending on the situation. Use whichever form of the CIRCLE command seems most appropriate or efficient to you.

Good luck!

DRAWING 2-1: APERTURE WHEEL

This drawing will give you practice drawing circles using the center point, radius method. Refer to the table below the drawing for radius sizes. With snap set at .25, some of the circles can be drawn by dragging and pointing. Other circles have radii that are not on a snap point. These circles can be drawn easily by typing in the radius.

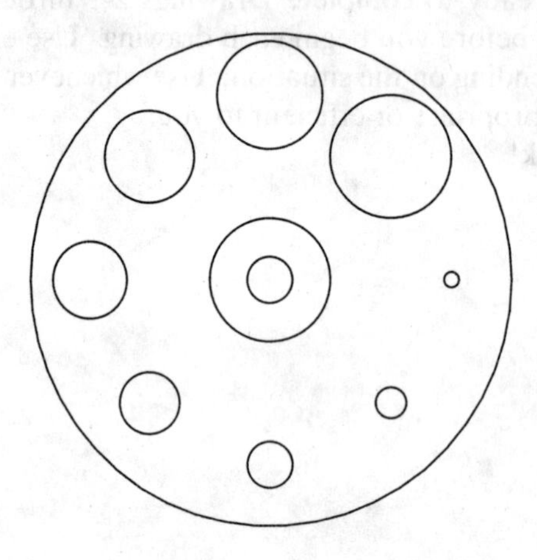

DRAWING SUGGESTIONS

GRID = .50
SNAP = .25

> A good sequence for doing this drawing would be to draw the outer circle first, followed by the two inner circles (h and c). These are all centered on the point (6.00,4.50). Then begin at circle a and work around clockwise, being sure to center each circle correctly.
> Notice that there are two circles c and two h. This simply indicates that the two circles having the same letter are the same size.
> Remember, you may type any value you like and AutoCAD will give you a precise graphic image, but you cannot always show the exact point you want with a pointing device. Often, it is more efficient to type a few values than to turn snap off or change its setting for a small number of objects.

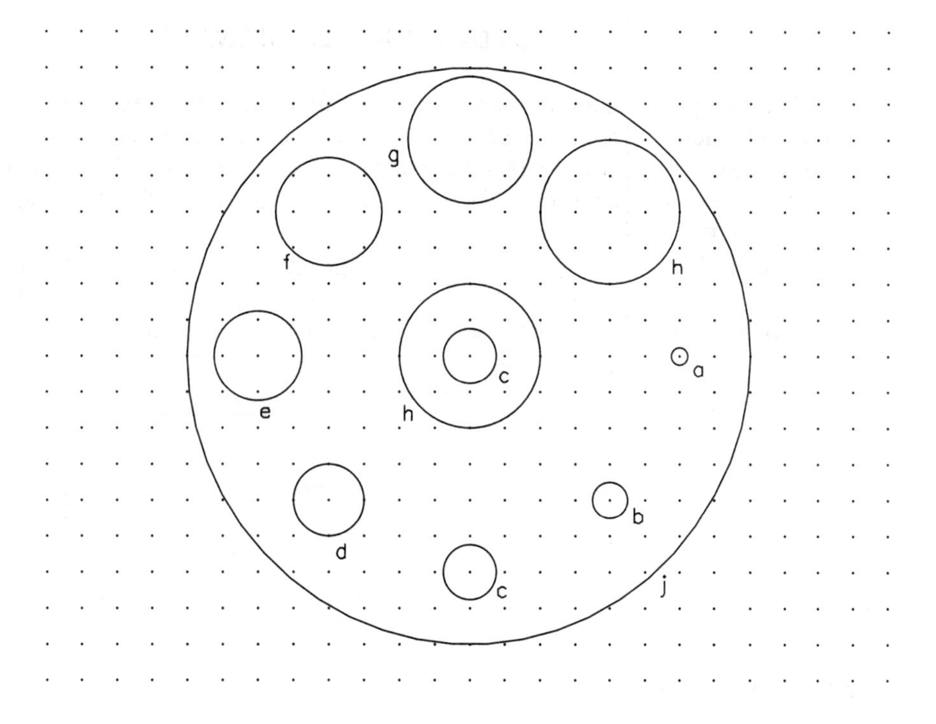

LETTER	a	b	c	d	e	f	g	h	j
RADIUS	.12	.25	.38	.50	.62	.75	.88	1.00	4.00

APERTURE WHEEL

Drawing 2-1

LAYER	0	CIRCLE	(C)
UNITS	2-PLACE DECIMAL	FRASE	(E)
GRID	.50	OOPS	
SNAP	.25	REDRAW	(R)

F1	F2	F6	F7	F8	F9
HELP	TEXT/GRAPHICS SCREEN	ABSOLUTE/OFF/POLAR COORDS	ON/OFF GRID	ON/OFF ORTHO	ON/OFF SNAP

DRAWING 2-2: ROLLER

This drawing will give you a chance to combine lines and circles and to use the center point, diameter method. It will also give you some experience with smaller objects, a denser grid, and a tighter snap spacing.

NOTE: Even though units are set to show only two decimal places, it is important to set the snap using three places (.125) so that the grid is on a multiple of the snap (.25 = 2 × .125). AutoCAD will show you rounded coordinate values, like .13, but will keep the graphics on target. Try setting snap to either .13 or .12 instead of .125, and you will see the problem for yourself.

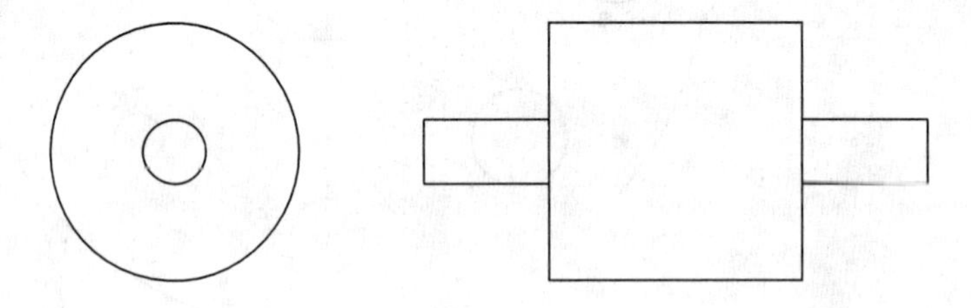

DRAWING SUGGESTIONS

GRID = .25
SNAP = .125

> The two views of the roller will appear fairly small on your screen, making the snap setting essential. Watch the coordinate display as you work and get used to the smaller range of motion.

> Choosing an efficient sequence will make this drawing much easier to do. Since the two views must line up properly, we suggest that you draw the front view first, with circles of diameter .25 and 1.00, and then use these circles to position the lines in the right side view.

> The circles in the front view should be centered in the neighborhood of (2.00,6.00). This will put the upper left-hand corner of the 1 × 1 square at around (5.50,6.50) .

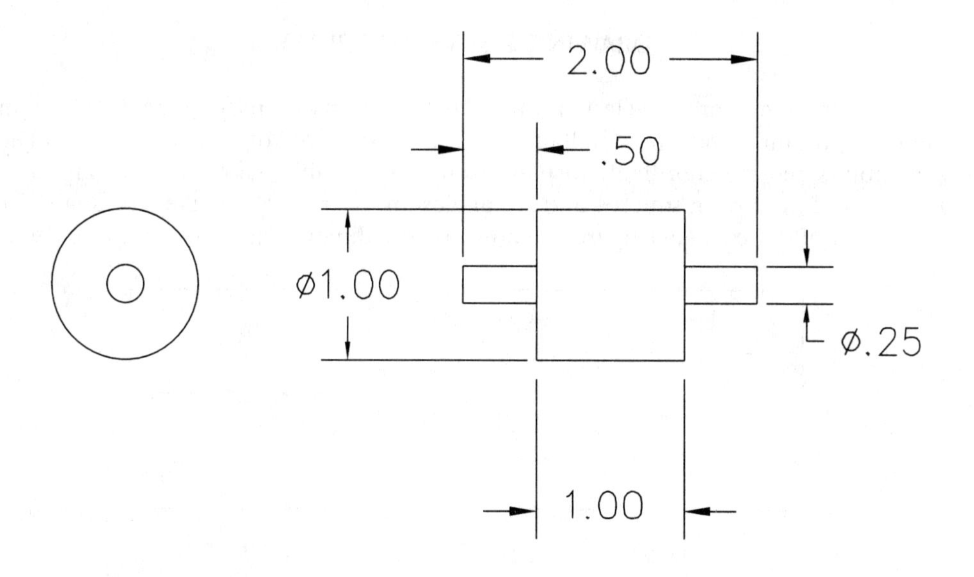

ROLLER

Drawing 2-2

LAYER	0	LINE	(L)
UNITS	2-PLACE DECIMAL	CIRCLE	(C)
GRID	.25	ERASE	(E)
SNAP	.125	OOPS	
		REDRAW	(R)

F1	F2	F6	F7	F8	F9
HELP	TEXT/GRAPHICS SCREEN	ABSOLUTE/OFF/POLAR COORDS	ON/OFF GRID	ON/OFF ORTHO	ON/OFF SNAP

DRAWING 2-3: SWITCH PLATE

This drawing is similar to the last one, but the dimensions are more difficult and a number of important points do not fall on the grid. It will give you practice using grid and snap points and the coordinate display. Refer to the table below the drawing for dimensions of the circles, squares, and rectangles inside the 7 × 10 outer rectangle. The placement of these smaller figures is shown by the dimensions on the drawing itself.

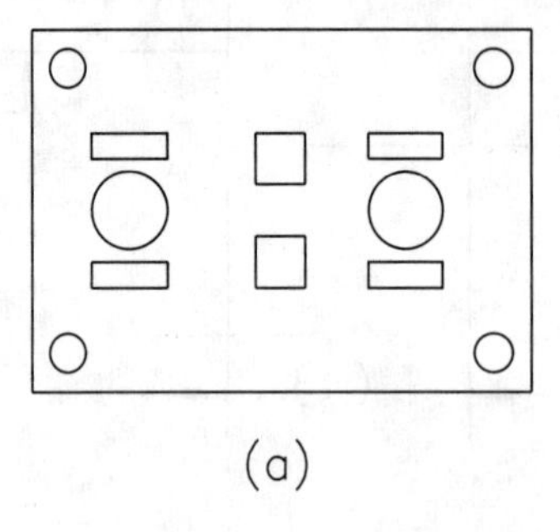

(a)

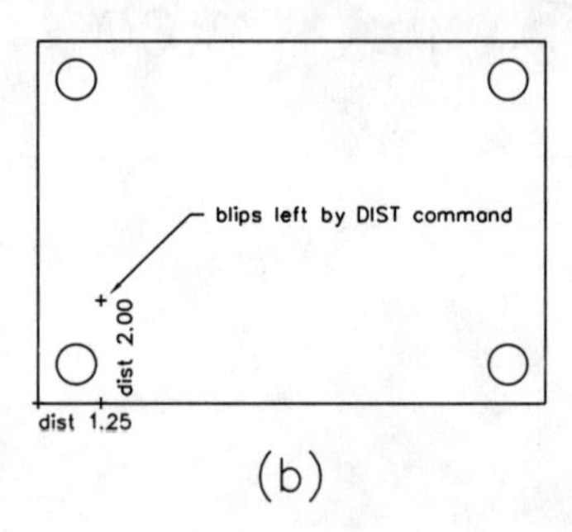

(b)

DRAWING SUGGESTIONS

GRID = .50
SNAP = .25

> Turn ortho on to do this drawing.
> A starting point in the neighborhood of (1,1) will keep you well positioned on the screen.

THE RECTANG COMMAND

The RECTANG command is a quick and easy way to draw rectangles. Type or select the command and then pick two points at opposite corners of the rectangle you want to draw. It is just like creating a selection window, but the window remains as a single entity called a polyline. Polylines are discussed in Chapter 10. Type "Rectang" or select the Rectangle tool from the Draw toolbar.

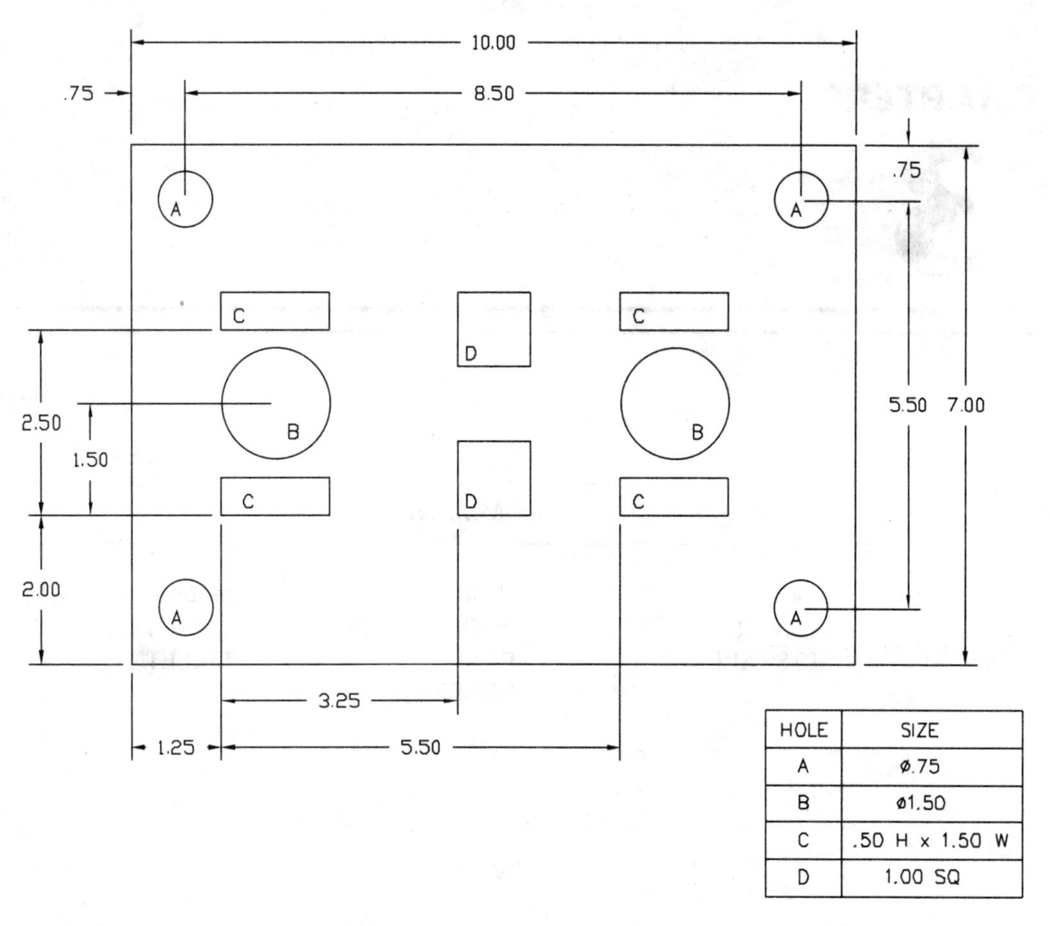

HOLE	SIZE
A	ø.75
B	ø1.50
C	.50 H x 1.50 W
D	1.00 SQ

SWITCH PLATE

Drawing 2–3

LAYER	0	LINE	(L)
UNITS	2-PLACE DECIMAL	CIRCLE	(C)
GRID	.50	ERASE	(E)
SNAP	.25	OOPS	
		REDRAW	(R)

F1	F2	F6	F7	F8	F9
HELP	TEXT/GRAPHICS SCREEN	ABSOLUTE/OFF/POLAR COORDS	ON/OFF GRID	ON/OFF ORTHO	ON/OFF SNAP

CHAPTER 3

COMMANDS

DATA	VIEW	MODIFY
LAYER	ZOOM	CHAMFER
LTSCALE	PAN	FILLET
	REGEN	

OVERVIEW

So far, all the drawings you have done have been on a single white layer called "0". In this chapter you will create and use three new layers, each with its own associated color and linetype.

You will also learn to FILLET and CHAMFER the corners of previously drawn objects, to magnify portions of a drawing using the ZOOM command, and to move between adjacent portions of a drawing with the PAN command.

TASKS

1. Create three new layers.
2. Assign colors to layers.
3. Assign linetypes to layers.
4. Change the current layer.
5. FILLET the corners of a square.
6. CHAMFER the corners of a square.

7. ZOOM in and out using Window, Previous, and All.
8. PAN to display another area of a drawing.
9. Use the Plot Preview feature.
10. Do Drawing 3-1 (“Mounting Plate”).
11. Do Drawing 3-2 (“Bushing”).
12. Do Drawing 3-3 (“Half Block”).

TASK 1: Creating New LAYERS

Layers allow you to treat specialized groups of entities in your drawing separately from other groups. For example, all of the dimensions in this book were drawn on a special dimension layer so that we could turn them on and off at will. We turned off the dimension layer in order to prepare the reference drawings for Chapters 1 through 7, which are shown without dimensions. When a layer is turned off, all the objects on that layer become invisible, though they are still part of the drawing database and can be recalled at any time.

It is common to put dimensions on a separate layer, and there are many other uses of layers as well. Fundamentally, layers are used to separate colors and linetypes, and these in turn take on special significance depending on the drawing application. It is standard drafting practice, for example, to use small, evenly spaced dashes to represent objects or edges that would in reality be hidden from view. On a CAD system with a color monitor, these hidden lines can also be given their own color to make it easy for the designer to remember what layer he or she is working on.

In this book we will use a simple and practical layering system, most of which will be presented in this chapter. You should remember that there are many other systems in use, and many other possibilities. AutoCAD allows as many as 256 different colors and as many layers as you like.

You should also be aware that linetypes and colors are not restricted to being associated with layers. It is possible to mix linetypes and colors on a single layer. While this may be useful for certain applications, we do not recommend it at this point.

> Begin a new drawing and use the No Prototype check box to ensure that you are using the same defaults as those used in this chapter.

The Layer Control Dialogue Box (DDLMODES)

In this instance, we will introduce the dialogue box procedures first because they are more efficient than the command line methods. The main advantage of using the dialogue box is that a table of layers will be displayed in front of you as you make changes, and you will be able to make several changes at once.

> Select “Layers . . .” from the Data pull down menu.

This will call up the Layer Control dialogue box illustrated in *Figure 3-1.* At the top left you will see that the current layer is 0. Below that you will see a box that lists layers defined in the current drawing. In our case, “0” is the only defined layer. It is on, uses the color white, and the continuous linetype.

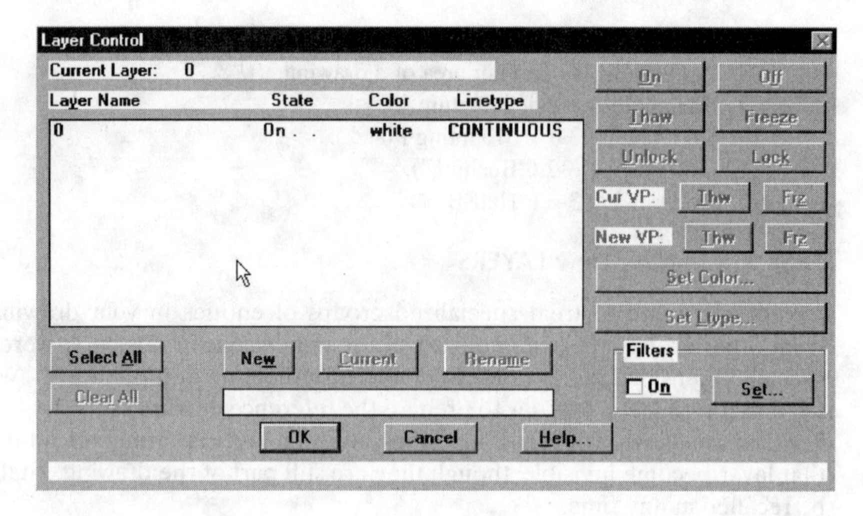

Figure 3-1

Now we will create three new layers. You should see a blinking line in the edit box at the bottom. This shows that AutoCAD is ready to accept typed input into this box.

> Type “1”. There is no need to press enter when using an edit box.

> Click on “New”, just above the edit box.

The newly defined layer, layer 1, will be entered immediately in the layer name box. You will see that it is currently defined with identical characteristics as layer 0. We will alter these momentarily. First, though, define two more layers.

Layer names may be up to thirty-one characters long, although only eighteen characters will be displayed in the layer name list. We have chosen single-digit numbers as layer names because they are easy to type and because we can match them to AutoCAD’s color numbering sequence.

You can also create several new layers at once by typing names in the edit box separated by commas. Try it.

> Type “2,3” in the edit box and click on “New” to create layers 2 and 3.

At this point, your layer name list should look like this:

LAYER NAME	STATE	COLOR	LINETYPE
0	On. .	white	CONTINUOUS
1	On. .	white	CONTINUOUS
2	On. .	white	CONTINUOUS
3	On. .	white	CONTINUOUS

NOTE: If you do not see the layer in the layer list, it will not be created. In order to create a new layer, you must click on “New”. A common error is to type names in the edit box and then click on “OK”. This will accomplish nothing. Also, if you cancel the dialogue box, all changes will be canceled including the creation of new layers.

TASK 2: Assigning Colors to LAYERS

We now have four layers, but they are all pretty much the same. Obviously, we have more changes to make before our new layers will have useful identities.

It is common practice to leave layer 0 defined the way it is. We will begin our changes on layer 1.

> (If for any reason you have left the Layer Control dialogue box, reenter it by clicking on "Layers . . ." from the Data pull down menu.)

Before you can change the qualities of a layer, you must have it selected in the layer name box. Until you do, all of the boxes on the right, including color and linetype, will be grayed out (inaccessible).

> Move the cursor arrow anywhere on the layer 1 line and click to select layer 1.

You will know the layer is selected when the line changes color.

> Click on "Set Color . . .".

This will call up the Select Color dialogue box illustrated in *Figure 3-2.* There are nine standard colors at the top, followed by gray shades and a palette of colors.

AutoCAD can display up to 256 different color shades. The nine standard colors shown at the top of the box are numbered one through nine and are the same for all color monitors. Color numbers 10–249 are shown in the full color palette and numbers 250–255 are the gray shades shown in between.

> Select the red box or type "1" or "red" in the edit box.

All of these methods will have the same result.

> Click on "OK".

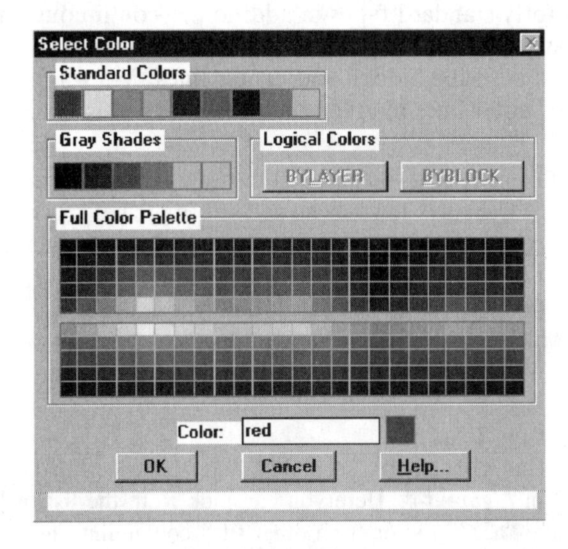

Figure 3-2

> Type "1" or "red" or select the red box on the left. If you type, pressing enter will complete the process and close the box. If you select, you will need to click on "OK" to complete the process.

You will now see that layer 1 is defined with the color red in the layer name list box.

Next we will assign the color yellow to layer 2. In order to do this, we need to select layer 2 in the box and de-select layer 1.

> Click on layer 2.

The layer 2 line should now be highlighted. Notice that layer 1 is still highlighted. This would allow you to set the layer qualities of more than one layer at once if the layers' qualities were to be the same.

> Click on the layer 1 line to deselect layer 1.

This line should no longer be highlighted.

> Click on "Set Color . . ." and assign the color yellow (color #2) to layer 2.

> Select layer 3, and de-select layer 2.

> Click on "Set Color . . ." again and assign the color green (color #3) to layer 3.

Look at the layer list. It should look like this:

LAYER NAME	STATE	COLOR	LINETYPE
0	On. .	white	CONTINUOUS
1	On. .	red	CONTINUOUS
2	On. .	yellow	CONTINUOUS
3	On. .	green	CONTINUOUS

TASK 3: Assigning Linetypes

AutoCAD has a standard library of linetypes that can be assigned easily to layers. There are forty standard types in addition to continuous lines. If you do not assign a linetype, AutoCAD will assume you want continuous lines. In addition to continuous lines we will be using hidden and center lines. We will put hidden lines in yellow on layer 2 and center lines in green on layer 3.

> (If for any reason you have left the Layer Control dialogue box, reenter it by clicking open the Data pull down menu and selecting "Layers . . .").

> Select layer 2 in the layer name box. Make sure other layers are not highlighted.

> Click on "Set Ltype . . .".

This will call up the Select Linetype dialogue box illustrated in *Figure 3-3.* The box which contains a list of loaded linetypes currently only shows the continuous linetype. We can fix this by selecting the "Load . . ." button at the bottom of the dialogue box.

> Click on "Load . . .".

This will call up a second dialogue box, the Load or Reload Linetypes box illustrated in *Figure 3-4.* Here you can pick from the list of linetypes available from the standard "acad" file or from other files containing linetypes, if there are any on your system. You also have the option of loading all linetypes from any given file at once.

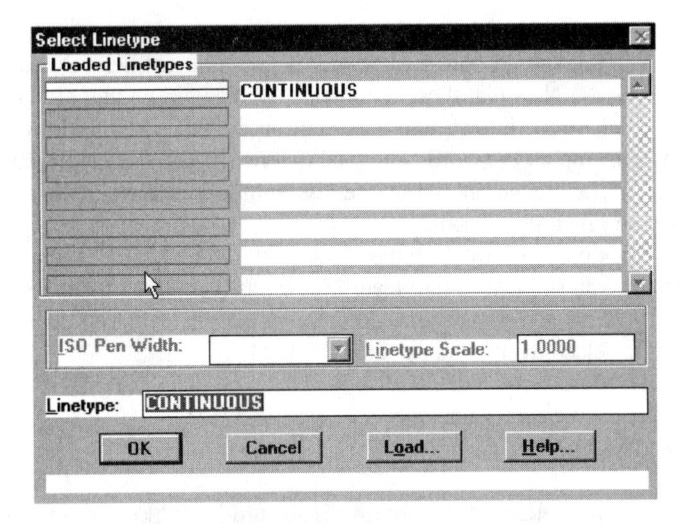

Figure 3-3

Load or Reload Linetypes
File...
acad
Available linetypes
BORDER
BORDER2
BORDERX2
CENTER
CENTER2
CENTERX2
DASHDOT
DASHDOT2
DASHDOTX2
DASHED
DASHED2
Select All
Clear All
OK
Cancel
Help...

Figure 3-4

The linetypes are then defined in your drawing and you can assign a new linetype to a layer at any time. This makes things easier. It does, however, use up more memory.

For our purposes, we will load only the hidden and center linetypes we are going to be using.

> Click on the word "Center" or anywhere on the center line.

The complete line for this linetype should now be highlighted.

> Move the cursor arrow to the down arrow on the scroll box at the right and scroll down until you see the "Hidden" linetype on the list.

> Click on the word "Hidden" or anywhere on the hidden line.

The hidden linetype should also be highlighted now.

> Click on "OK" to complete the loading process.

You should now see the center and hidden linetypes added to the list of loaded linetypes. Now that these are loaded, we can assign them to layers.

> Click on the hidden line illustrated in the box. You will notice that you can't select by clicking on the word "Hidden" this time. You have to select the line itself.

The hidden line should be highlighted in white and the word "HIDDEN" entered in the Linetype edit box.

> Click on "OK" to close the box.

You should see that layer 2 now has the hidden linetype.

Next assign the center linetype to layer 3.

> De-select layer 2 and select layer 3.

> Click on "Set Ltype . . .".

> Select the center linetype.

> Click on "OK".

Examine your layer list again. It should look like the one shown next. If not, use the other dialogue boxes to fix it.

LAYER NAME	STATE	COLOR	LINETYPE
0	On. .	white	CONTINUOUS
1	On. .	red	CONTINUOUS
2	On. .	yellow	HIDDEN
3	On. .	green	CENTER

TASK 4: Changing the Current LAYER

In order to draw new entities on a layer, you must make it the currently active layer. Previously drawn objects on other layers will also be visible and will be plotted if that layer is turned on, but new objects will go on the current layer.

> (If for any reason you have left the Layer Control dialogue box, reenter it by opening the Data pull down menu and selecting "Layers . . .").

> Select layer 1.

All other layers must be de-selected or else the "Current" box will be inaccessible.

> Click on "Current".

The current layer indicator at the top left of the dialogue box will show layer 1.

> When layer 1 and the current box are both highlighted, click on "OK".

At this point we suggest that you try drawing some lines to see that you are, in fact, on layer 1 and drawing in red, continuous lines.

When you are satisfied with the red lines you have drawn, go into the Layer Control dialogue box again (click twice on "Data") and set the current layer to 2. Now draw more lines and see that they are "hidden" yellow lines.

Set layer 3 as the current layer and draw some green center lines.

Finally, set layer 1 as the current layer before moving on.

TASK 5: Editing Corners Using FILLET

Now that you have a variety of linetypes to use, you can begin to do some more realistic mechanical drawings. All you will need is the ability to create filleted (rounded) and chamfered (cut) corners. The two work similarly, and AutoCAD makes them easy. Fillets may also be created between circles and arcs, but the most common usage is the type of situation demonstrated here.

> Erase any lines left on the screen from previous exercises.
> If you have not already done so, set layer 1 as the current layer.
> Draw a 5 × 5 square on your screen, as in *Figure 3-5*. We will use this figure to prac-

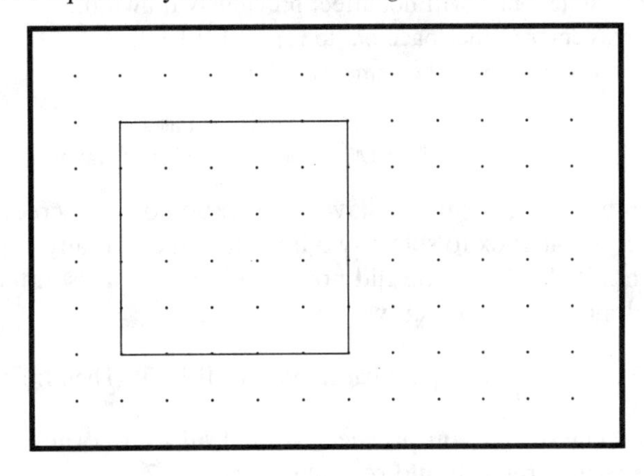

Figure 3-5

tice fillets and chamfers. Exact coordinates and lengths are not significant.
> Type "Fillet" or open the Modify toolbar and select the Fillet tool from the Feature flyout, as shown in *Figure 3-6*.

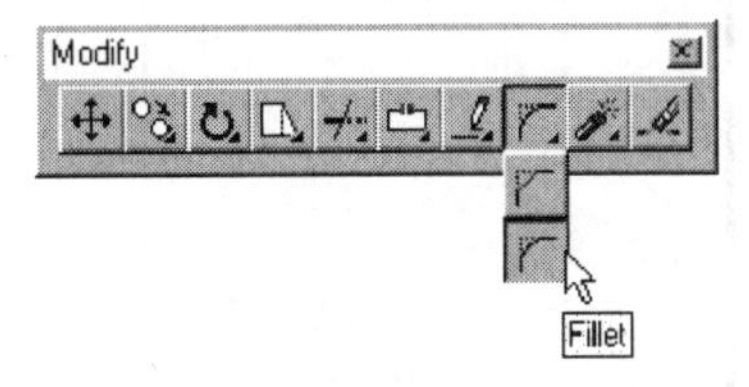

Figure 3-6

A prompt with options will appear in the command area and you will also see these options listed on the screen menu. Whenever you enter a command, options will be listed on the screen menu and can be selected if the screen menu is visible. The prompt is as follows:

(TRIM mode) Current fillet radius = 0.0000
Polyline/Radius/Trim<Select first object>:

The first thing you must do is determine the degree of rounding you want. Since fillets are really arcs, they can be defined by a radius.

> Type "r" or select "Radius" from the screen menu.

AutoCAD prompts:

```
Enter fillet radius <0.0000>:
```

You may have to open the text window (F2) to the complete prompt. The default is 0 because no fillet radius has been defined for this drawing yet. You can use a 0 fillet radius to connect two lines at a corner or, more commonly, you can define a fillet radius by typing a value or showing two points that define the radius length.

> Type ".5" or show two points .5 units apart.

You have set .5 as the standard fillet radius for this drawing. You can change it at any time, but it will not affect previously drawn fillets.

> Press enter or the space bar to repeat FILLET.

The prompt is the same as before:

```
(TRIM mode) Current fillet radius = 0.5000
Polyline/Radius/Trim<Select first object>:
```

You will notice that you have the pickbox on the screen now.

Use the pickbox to select two lines that meet at any corner of your square.

Behold! A fillet! You did not even have to press enter. AutoCAD knows that you are done after selecting two lines.

> Press enter or the space bar to repeat FILLET. Then fillet another corner.

We suggest that you proceed to fillet all four corners of the square. When you are done, your screen should resemble *Figure 3-7*.

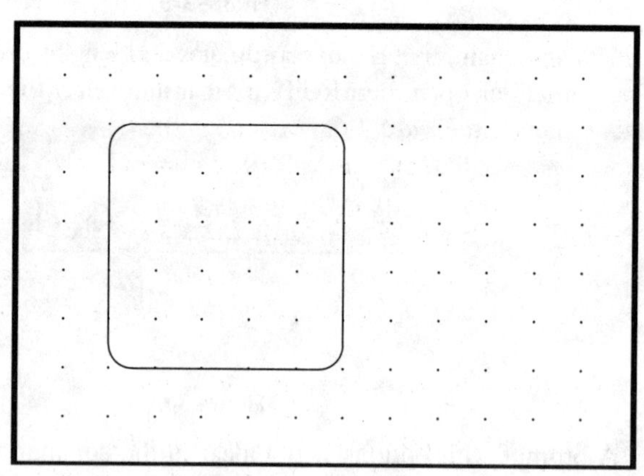

Figure 3-7

TASK 6: Editing Corners with CHAMFER

The CHAMFER command sequence is almost identical to the FILLET command, with the exception that chamfers may be uneven. That is, you may cut back farther on one side of a corner than the other. To do this, you must give AutoCAD two distances instead of one.

> Prepare for this exercise by undoing all your fillets with the U command as many times as necessary.
> Type "Chamfer" or select the Chamfer tool from the Feature flyout on the Modify toolbar, as shown in *Figure 3-8.*

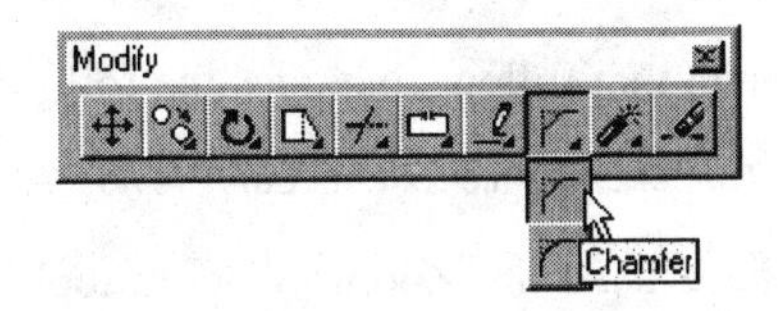

Figure 3-8

AutoCAD prompts:

(TRIM mode) Current chamfer Dist1 = 0.0000, Dist2 = 0.0000
Polyline/Distance/Angle/Trim/Method/<Select first line>:

> Type "d" or select "Distance".
The next prompt will be:

Enter first chamfer distance <0.0000>:

The present default is 0 because we have yet to define a chamfer distance for this drawing.
> Type ".25".
AutoCAD asks for another distance with a prompt like this:

Enter second chamfer distance <0.25>:

The first distance has become the default and most of the time it will be used. If you want an asymmetric chamfer, enter a different value for the second distance.
> Press enter to accept the default, making the chamfer distances symmetrical.
> Press enter to repeat the CHAMFER command.
> Answer the prompt by pointing to a line this time.
> Point to a second line, perpendicular to the first.
You should now have a neat chamfer on your square.
> Repeat CHAMFER and continue chamfering the other three corners of the square.

TASK 7: ZOOMing Window, Previous, and All

The capacity to zoom in and out of a drawing is one of the more impressive benefits of working on a CAD system. When drawings get complex, it often becomes necessary to work in detail on small portions of the drawing space. Especially with a small monitor, the only way to do this is by making the detailed area larger on the screen. This is easily done with the ZOOM command. You should have a square with chamfered corners on your screen from the previous exercise. If not, a simple square will do just as well, and you should draw one now.

> Type "z" or select the Zoom Window tool, as illustrated in *Figure 3-9.*

Figure 3-9

The prompt that follows includes the following options:

All/Center/Dynamic/Extents/Left/Previous/Vmax/Window/<Scale(X/XP)>:

If you have used the Zoom Window tool, the Window option will be entered automatically.

We are interested, for now, in All, Previous, and Window, which we will explore in reverse order. See the AutoCAD Command Reference for further information.

As in ERASE and other edit commands, you can force a window selection by typing "w" or selecting "Window". However, this is unnecessary. The windowing action is automatically initiated if you pick a point on the screen after entering ZOOM.

> Pick a point just below and to the left of the lower left-hand corner of your square (point 1 in *Figure 3-10*).

AutoCAD asks for another point:

Other corner:

You are being asked to define a window, just as in the ERASE command. This window will be the basis for what AutoCAD displays next. Since you are not going to make a window that exactly conforms to the screen size and shape, AutoCAD will interpret the window this way: Everything in the window will be shown, plus whatever additional area is needed to fill the screen. The center of the window will become the center of the new display.

> Pick a second point near the center of your square (point 2 in the figure).

The lower left corner of the square should now appear enlarged on your screen, as shown in *Figure 3-11*.

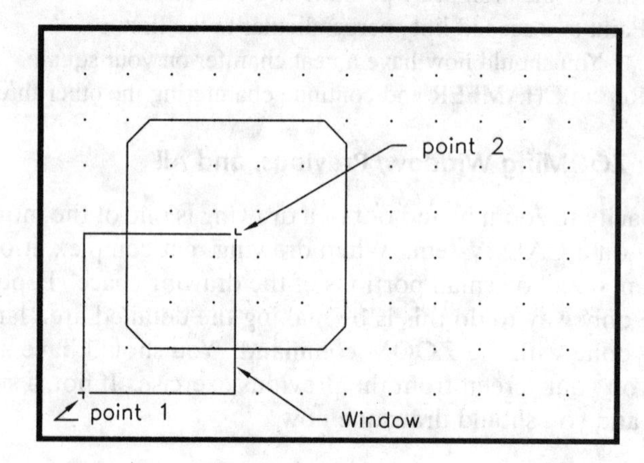

Figure 3-10

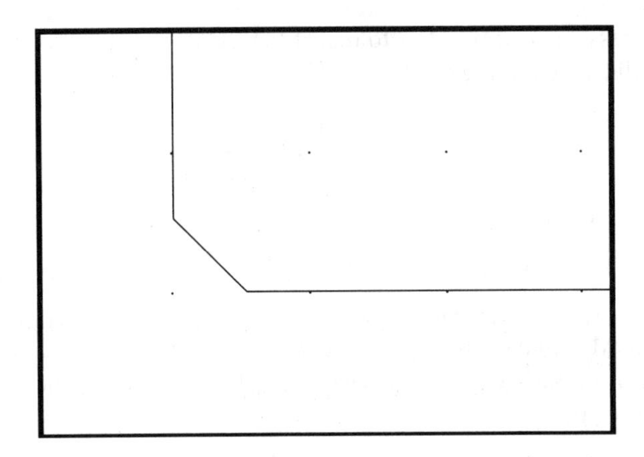

Figure 3-11

> Using the same method, try zooming up further on the chamfered corner of the square. If snap is on, you may need to turn it off (F9).

Remember that you can repeat the ZOOM command by pressing enter or the space bar.

At this point, most people cannot resist seeing how much magnification they can get by zooming repeatedly on the same corner or angle of a chamfer. Go ahead. After a couple of zooms the angle will not appear to change. An angle is the same angle no matter how close you get to it, but what happens to the spacing of the grid and snap as you move in?

When you are through experimenting with window zooming, try zooming out to the previous display.

> Press enter to repeat the ZOOM command. Do not use any of the ZOOM tools.

> Type "p".

You should now see your previous display.

AutoCAD keeps track of your most recent displays. The exact number of displays it stores depends on the version you are using. Release 13 remembers ten previous displays.

> ZOOM "Previous" as many times as you can until you get a message that says:

No previous display saved.

ZOOM ALL

ZOOM All zooms out to display the whole drawing. It is useful when you have been working in a number of small areas of a drawing and are ready to view the whole scene. You do not want to have to wade through previous displays to find your way back. ZOOM All will take you there in one jump.

In order to see it work, you should be zoomed in on a portion of your display before executing ZOOM All.

> Press enter or type "z" to repeat the ZOOM command, or select the Zoom All tool, as illustrated in *Figure 3-12*.

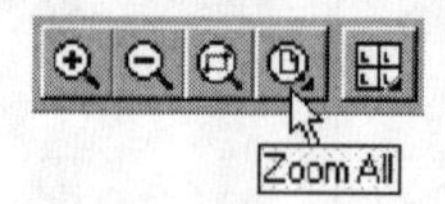

Figure 3-12

The Zoom All tool automatically enters the "All" option. If you have entered the ZOOM command by typing you will have the following additional step.
> If necessary, type "a" or select "All" from the screen menu.
There you have it.

NOTE: The Zoom All tool has a flyout that includes tools for all of the ZOOM command options.

ZOOM Tools

The other ZOOM tools, Zoom In and Zoom Out, are easy to use but provide less flexible movement. Zoom In always magnifies the image by a factor of 2. Zoom Out reverses this by using a constant factor of 0.5. Try these if you like. At any time, you can return to the whole drawing by using the Zoom All tool.

TASK 8: Moving the Display Area with PAN

As soon as you start to use ZOOM, you are likely to need PAN as well. While ZOOM allows you to magnify portions of your drawing, PAN allows you to shift the area you are viewing in any direction.

In Windows you have the option of panning with the scrollbars at the edges of the drawing area, but the PAN command allows more precision and flexibility.

> Type "p" or select the Pan tool from the Standard toolbar (illustrated in *Figure 3-13*).

Figure 3-13

AutoCAD will prompt you to show a displacement:

```
Displacement:
```

Imagine that your complete drawing is hidden somewhere behind your monitor, and that the display area is now functioning like a microscope with the lens focused on one portion. If the ZOOM command increases the magnification of the lens, then PAN moves the drawing like a slide under the lens, in any direction you want.

To move the drawing, you will indicate a displacement by picking two points on the screen. The line between them will serve as a vector, showing the distance and direction you want to PAN. Notice that the objects on your screen will move in the direction you indicate; if your vector moves to the right, so will the objects.

> Pick point 1 to begin your displacement vector, as shown in *Figure 3-14*.

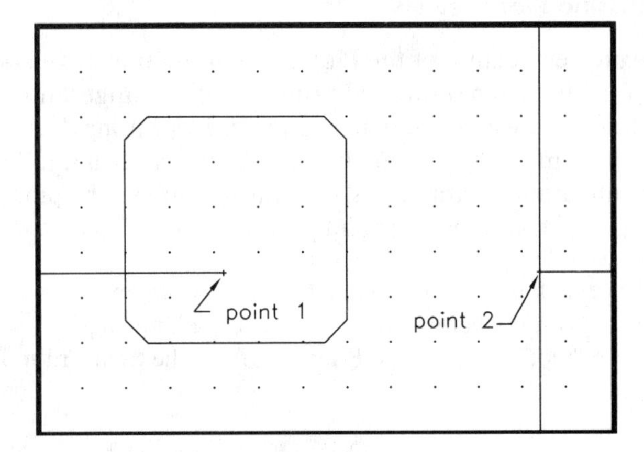

Figure 3-14

You will be prompted for another point:

Second point:

> Pick point 2 to the right of your first point.

As soon as you have shown AutoCAD the second point, objects on the screen will shift to the right, as in *Figure 3-15*.

> Press enter or the space bar to repeat the PAN command.

> Indicate a displacement to the left, moving objects back near their previous positions.

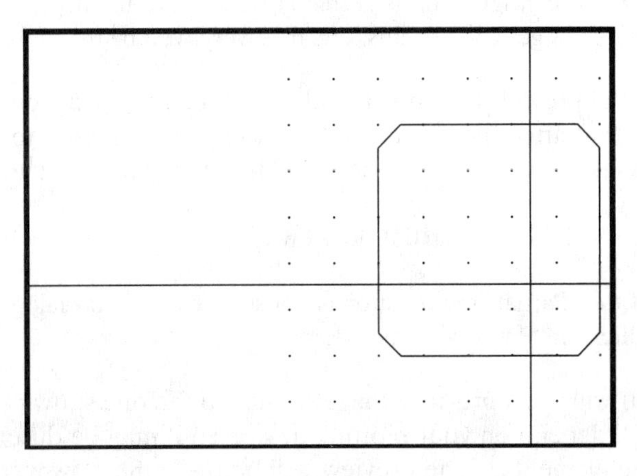

Figure 3-15

Experiment with the PAN command, moving objects up, down, left, right, and diagonally. You should also compare the effect of using the scroll bars with the PAN command. What function key would make it impossible to PAN diagonally? Hint: The name of the function begins and ends with an "o" and the F-key that turns it on and off begins with an "F" and ends with an "8".

TASK 9: Using Plot Preview

The plot preview feature of the Plot Configuration dialogue box is an essential tool in carrying out efficient plotting and printing. Plot configuration is complex and the odds are good that you will waste time and paper by printing drawings directly without first previewing them on the screen. In this task we are still significantly limited in our use of plot configuration parameters, but learning to use the plot preview will make all of your future work with plotting and printing more effective.

> To begin this task, you should have a drawing on your screen ready to preview. Any drawing will do, but we will illustrate using Drawing 3-2.
> Type "plot" or select the Print tool from the Standard toolbar as illustrated in *Figure 3-16.*

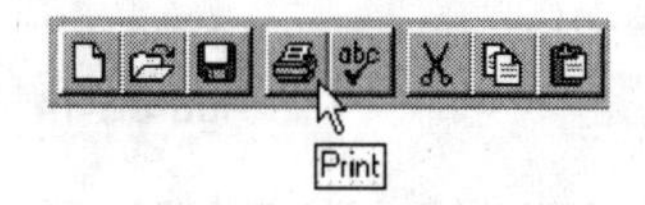

Figure 3-16

This will open the Plot Configuration dialogue box. Plot Preview is at the lower right of the dialogue box. There are two types of preview: "Partial" and "Full". A partial preview will show you an outline of the effective plotting area in relation to the paper size, but will not show an image of the plotted drawing.

NOTE: We will address paper sizes in the next chapter. For now, we will assume that your plot configuration is correctly matched to the paper in your printer or plotter. If you do not get good results, this may very well be the problem.

Partial previews are quick and should be accessed frequently as you set plot parameters that affect how paper will be used and oriented to print your drawing. Full previews take longer and may be saved for when you think you have gotten everything right.

We will look at a partial preview first.

> If the "Partial" radio button is not selected in your dialogue box, select it now.
> Click on "Preview . . .".

You will see a preview image similar to the one shown in *Figure 3-17.* The exact image will depend on your plotting device, so it may be different from the one shown here. The elements of the preview will be the same, however. The red rectangle rep-

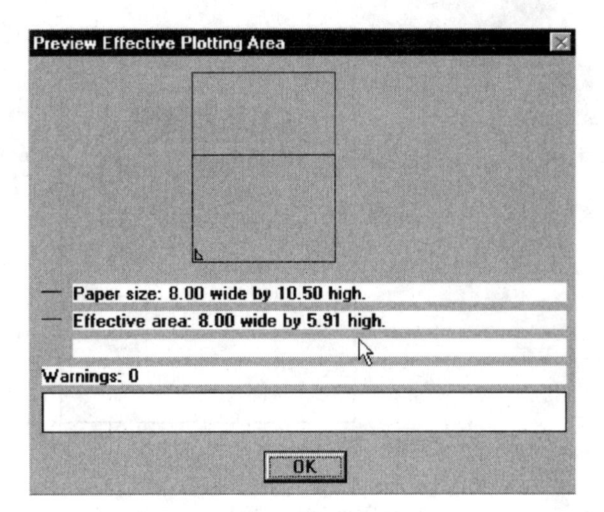

Figure 3-17

resents your drawing paper. It may be oriented in landscape (horizontally) or portrait (vertically). Typically, printers are in portrait and plotters are in landscape.

The blue rectangle inside the red one illustrates the effective plotting area. It is quite possible that the blue and red rectangles will overlap. The blue represents the size and shape of the area that AutoCAD can actually use given the shape of the drawing in relation to the size and orientation of the paper. The effective area is dependent on many things, as you will see. We will leave it as is for now and return later when you begin to alter the plot configuration.

In one corner of the red rectangle, you will see a small triangle. This is the rotation icon. It shows the corner of the plotting area where the plotter will begin plotting (the origin).

> Click on "OK" to exit the preview box.
> Click on the "Full" radio button to switch to a full preview.
> Click on "Preview . . .".

The dialogue box will disappear temporarily and you will see a preview similar to the one in *Figure 3-18.* We have used Drawing 3-2 to illustrate. You will see whatever drawing you are previewing, with orientation and placement depending on your plotting device.

The small Plot Preview box in the middle can be moved aside like any dialogue box. Just move the cursor to the gray title area, press the pick button, and hold it down as you move the box.

The pan and zoom feature allows you to look closely at small areas in the drawing (to check text or dimensions, for example). Panning and zooming in the preview has no effect on the plot parameters.

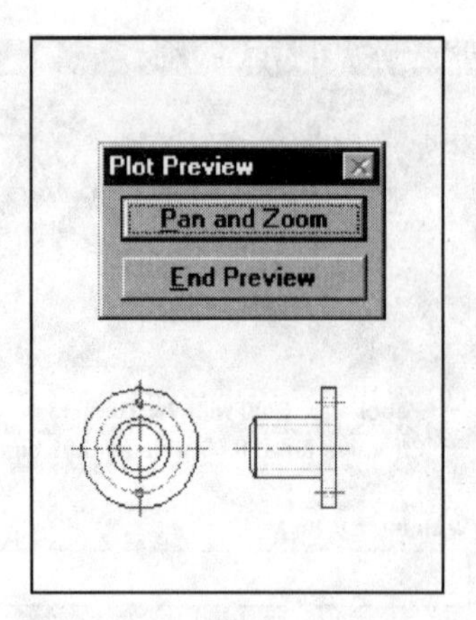

Figure 3-18

> Click on "Pan and Zoom".

Selecting "Pan and Zoom" will cause the box to disappear and be replaced by a rectangle with an X in the middle. The box shows the portion of the drawing that will be shown when the pan/zoom is executed. With the X showing, you are in "pan" mode. You can move the box to any portion of the drawing you want to examine. Pressing enter while in pan mode will cause AutoCAD to show you an image of the portion of your drawing that is within the pan box.

> Press the pick button once. (Do not press enter or the enter button on your cursor yet.)

When you press the pick button once, the X will be replaced by an arrow at the right side of the box. Then you will be able to shrink or stretch the box by moving the cursor left or right. This is the "zoom" mode which allows you to focus on larger or smaller segments of the drawing. Notice that you cannot move the box to the left. You can only cause it to expand or contract to the right. To move to the left, you need to press the pick button again to get back into pan mode. To get the precise area you want, you may need to switch back and forth between pan and zoom several times. When you have moved and stretched the box so that it windows the area you want to see, press the second button on your cursor or press enter. This will execute the magnification and you will see an enlarged image. We suggest that you try it on whatever drawing you are looking at.

> Pan and zoom to window an area of your drawing that contains objects.

> Press enter, the space bar, or the enter button on your pointing device.

You will see an enlarged image of that portion of your drawing.

> Click on "Zoom Previous" to return to the full drawing preview.

> Click on "End Preview" to exit the full preview.

This will bring you back to the Plot Configuration dialogue.

This ends our initial preview of your drawing. In ordinary practice, if everything looks right in the preview you would move on to plot or print your drawing now by clicking on "OK". In the chapters that follow, we will explore the rest of the plot configuration dialogue box and will use plot preview extensively as we change plot parameters. For now, get used to using plot preview. If things are not coming out quite the way you want, you will be able to fix them soon.

TASKS 10, 11, and 12

With layers, colors, linetypes, fillets, chamfers, zooming, and panning you are ready to do the drawings for Chapter 3.

Remember to set grid, snap, and units and to define layers before you begin each drawing. Use the ZOOM and PAN commands whenever you think they will help you to draw more efficiently. When you are ready to plot a drawing, take full advantage of the plot preview feature in the Plot Configuration dialogue box.

DRAWING 3-1: MOUNTING PLATE

This drawing will give you experience using center lines and chamfers. Since there are no hidden lines, you will have no need for layer 2, but we will continue to use the same numbering system for consistency. Draw the continuous lines in red on layer 1 and the center lines in green on layer 3.

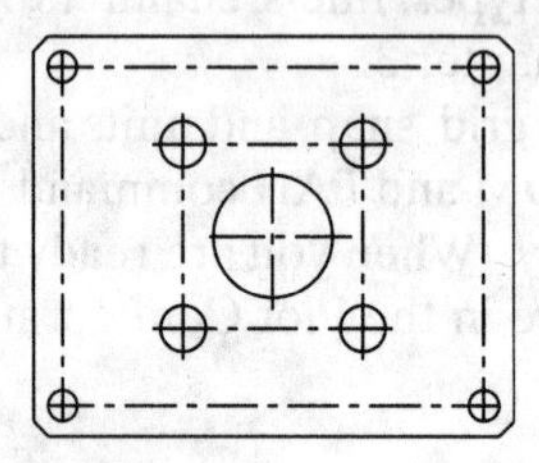

DRAWING SUGGESTIONS

GRID = .5
SNAP = .25
LTSCALE = .5

LTSCALE

The size of the individual dashes and spaces that make up center lines, hidden lines, and other linetypes is determined by a global setting called "LTSCALE". By default, it is set to a factor of 1.00. In smaller drawings, this setting will be too large and cause some of the shorter lines to appear continuous regardless of what layer they are on.

To remedy this, change LTSCALE as follows:

1. Type "ltscale" or select "Linetypes" and then "Global Linetype Scale" from the Options pull down menu.
2. Enter a value.

For the drawings in this chapter, use a setting of .50. See *Figure 3-19* for some examples of the effect of changing LTSCALE.

LTSCALE = 1.00
LTSCALE = .50
LTSCALE = .25

Figure 3-19

> Draw the chamfered rectangle and the nine circles on layer 1 first. Then, set current layer to 3 and draw the center lines.

NOTE: In manual drafting, it would be more common to draw the center lines first and use them to position the circles. Either order is fine, but be aware that what is standard practice in pencil and paper drafting may not be efficient or necessary on a CAD system.

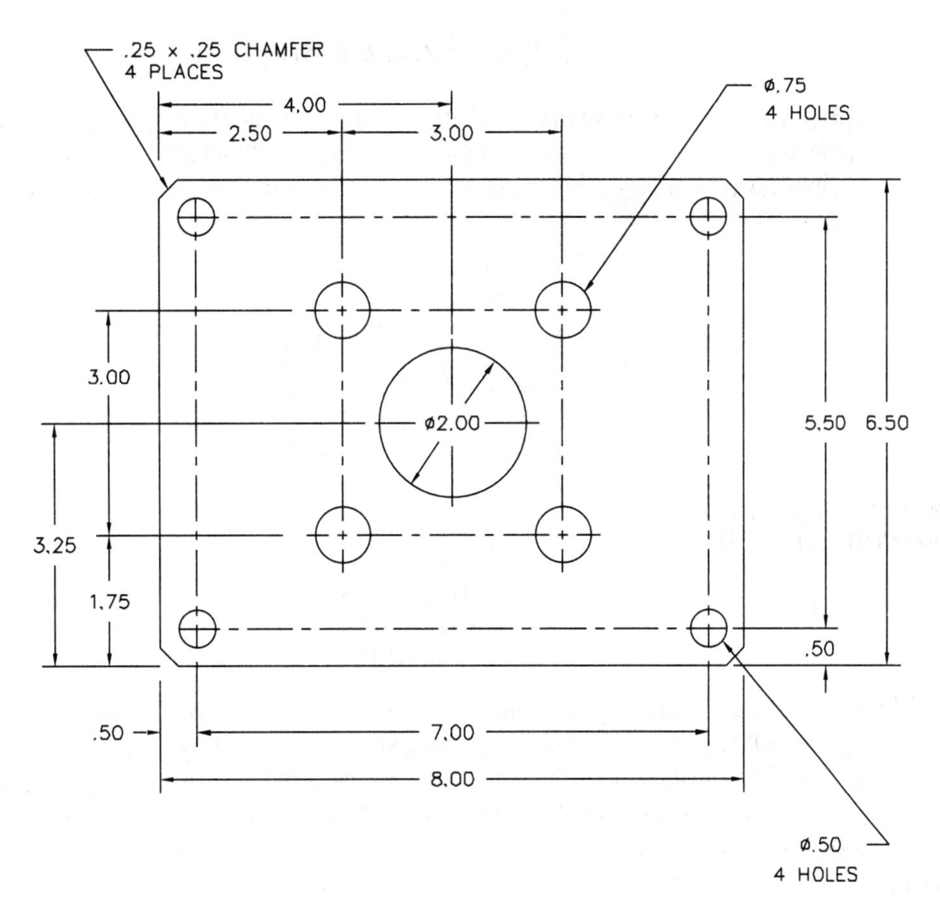

MOUNTING PLATE

Drawing 3–1

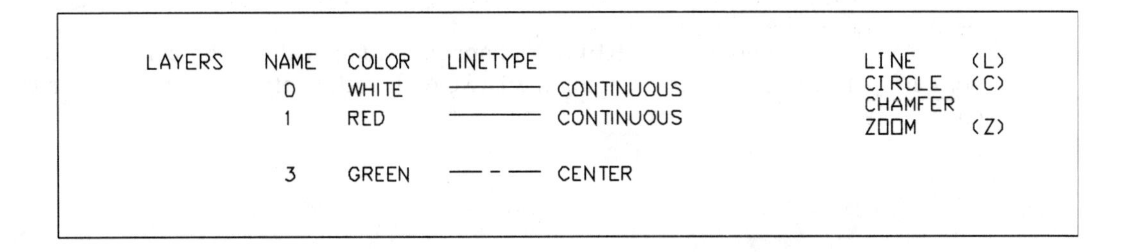

F1	F2	F6	F7	F8	F9
HELP	TEXT/GRAPHICS SCREEN	ABSOLUTE/OFF/POLAR COORDS	ON/OFF GRID	ON/OFF ORTHO	ON/OFF SNAP

DRAWING 3-2: BUSHING

This drawing will give you practice with chamfers, layers, and zooming. Notice that because of the smaller dimensions here, we have recommended a smaller LTSCALE setting.

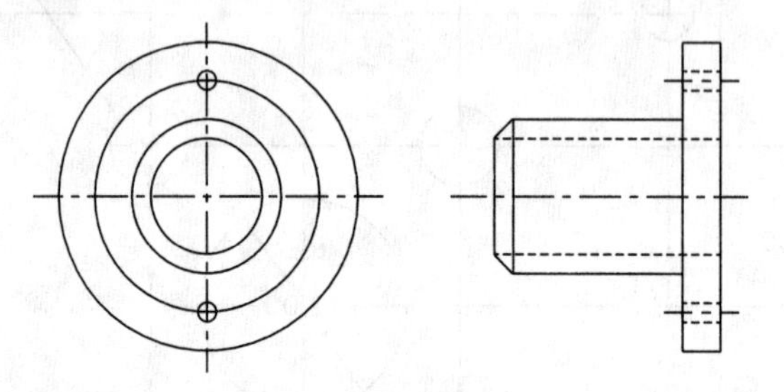

DRAWING SUGGESTIONS

GRID = .25
SNAP = .125
LTSCALE = .25

> Since this drawing will appear quite small on your screen, it would be a good idea to ZOOM in on the actual drawing space you are using, and use PAN if necessary.
> Notice that the two .25-diameter screw holes are 1.50 apart. This puts them squarely on grid points that you will have no trouble finding.

REGEN

When you zoom you may find that your circles turn into many-sided polygons. AutoCAD does this to save time. These time savings are not noticeable now, but when you get into larger drawings they become very significant. If you want to see a proper circle, type "Regen". This command will cause your drawing to be regenerated more precisely from the data you have given.

You may also notice that REGENs happen automatically when certain operations are performed, such as changing the LTSCALE setting after objects are already on the screen.

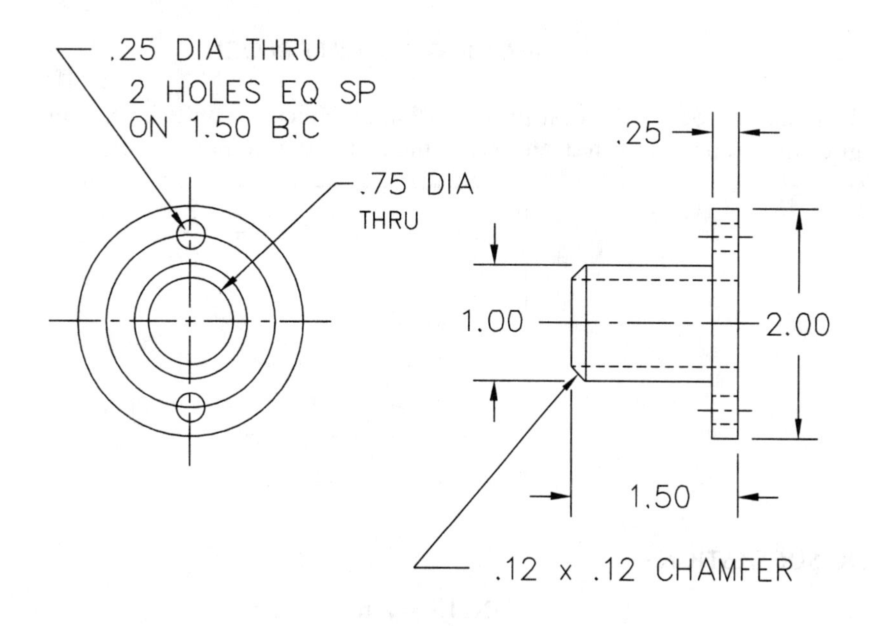

BUSHING
Drawing 3–2

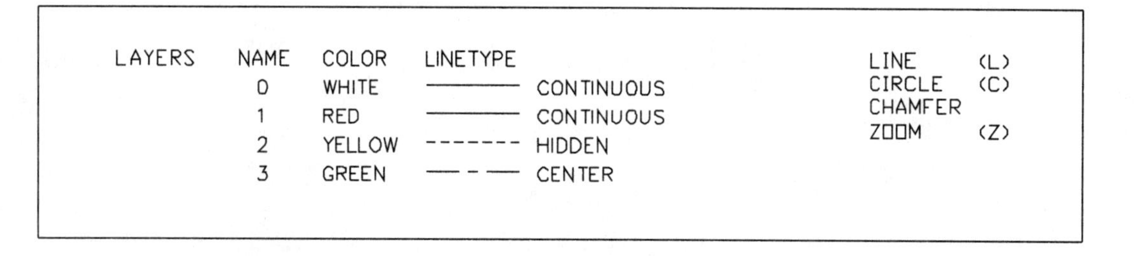

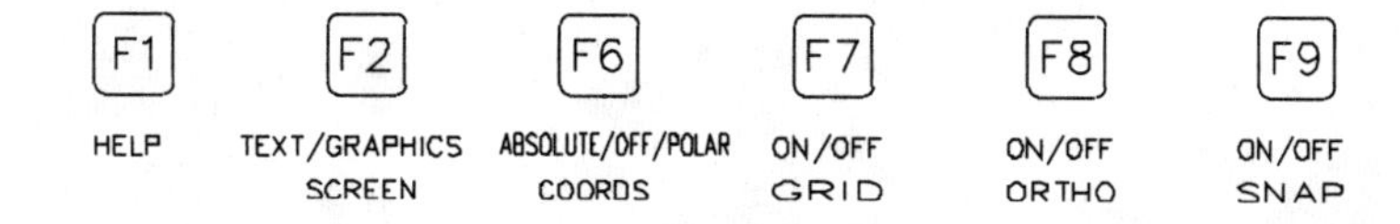

DRAWING 3-3: HALF BLOCK

This cinder block is the first project using architectural units in this book. Set units, grid, and snap as indicated, and everything will fall into place nicely.

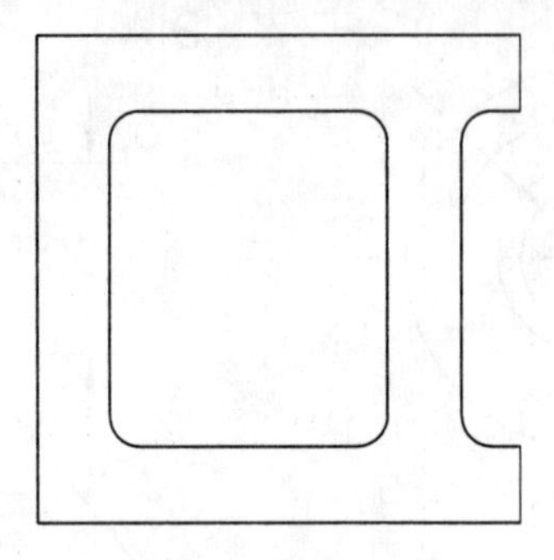

DRAWING SUGGESTIONS

UNITS = Architectural
smallest fraction = 4 (1/4″)
GRID = 1/4″
SNAP = 1/4″

> Start with the lower left corner of the block at the point (0′-1″,0′-1″) to keep the drawing well placed on the display.
> After drawing the outside of the block with the 5 1/2″ indentation on the right, use the DIST command to locate the inner rectangle 1 1/4″ in from each side.
> Set the FILLET radius to 1/2″ or .5. Notice that you can use decimal versions of fractions. The advantage is that they are easier to type.

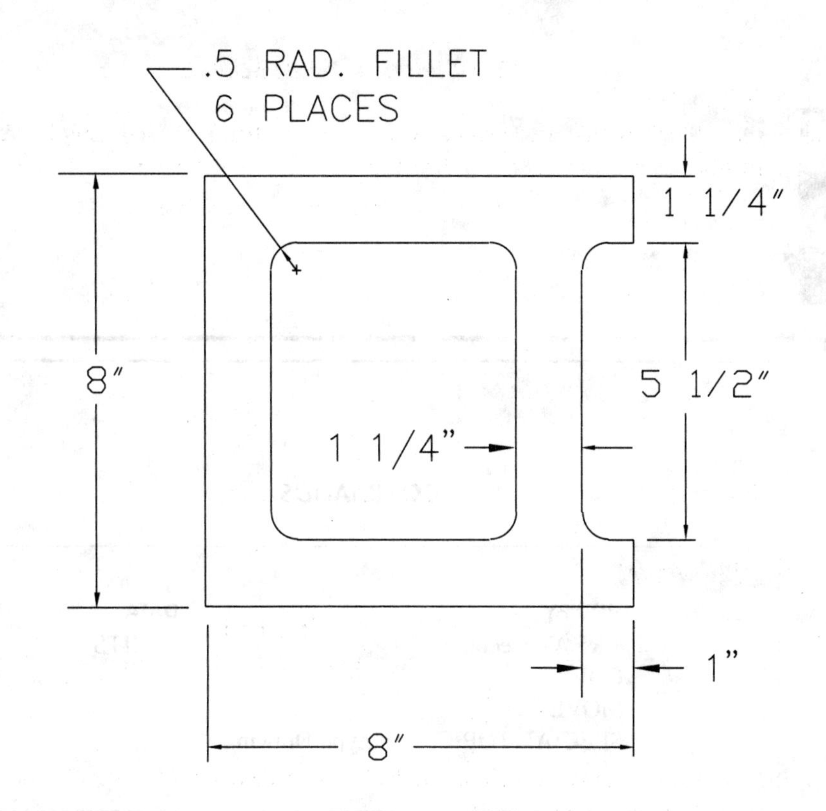

HALF BLOCK

Drawing 3–3

LAYERS	NAME	COLOR	LINETYPE	
	0	WHITE	———	CONTINUOUS
	1	RED	———	CONTINUOUS

LINE (L)
FILLET
ERASE (E)
ZOOM (Z)

F1	F2	F6	F7	F8	F9
HELP	TEXT/GRAPHICS SCREEN	ABSOLUTE/OFF/POLAR COORDS	ON/OFF GRID	ON/OFF ORTHO	ON/OFF SNAP

CHAPTER 4

COMMANDS

MODIFY	DATA
ARRAY (rectangular)	LIMITS
COPY	
MOVE	
SPECIAL TOPIC: Prototype Drawings	

OVERVIEW

In this chapter, you will learn some real timesavers. If you have grown tired of defining the same three layers, along with units, grid, snap, and ltscale for each new drawing, read on. You are about to learn how to use prototype drawings so that every time you begin a new drawing you will begin with whatever setup you want. In addition, you will learn to reshape the grid using the LIMITS command and to COPY and MOVE objects on the screen. We will begin with LIMITS, since we will want to change the limits as part of defining your first prototype.

TASKS

1. Change the shape of the grid using the LIMITS command.
2. Create a prototype drawing.
3. Select your drawing as the prototype.
4. MOVE an object in a drawing.

5. COPY an object in a drawing.
6. Create a rectangular ARRAY.
7. Change Plot Configuration parameters.
8. Do Drawing 4-1 ("Grill").
9. Do Drawing 4-2 ("Test Bracket").
10. Do Drawing 4-3 ("Floor Framing").

TASK 1: Setting LIMITS

You have changed the density of the screen grid many times, but always within the same 12 × 9 space, which basically represents an A-size sheet of paper. Now you will learn how to change the shape by setting new limits to emulate other sheet sizes or any other space you want to represent. First, a word about model space and paper space.

Model Space and Paper Space

"Model space" is an AutoCAD concept that refers to the imaginary space in which we create and edit objects. In model space, objects are always drawn full scale (1 screen unit = 1 unit of length in the real world). The alternative to model space is paper space, in which screen units represent units of length on a piece of drawing paper. Paper space is most useful in plotting multiple views of 3D drawings. In part 1 of this book, all of your work will be done in model space.

In this exercise, we will reshape our model space to emulate different drawing sheet sizes. You should be aware, however, that this is by no means a necessary practice. With AutoCAD you will be able to scale your drawing to fit any drawing sheet size when it comes time to plot. Model space limits should be determined by the size and shapes of objects in your drawing, not by the paper you are going to use when you plot. However, setting limits in paper space is done in exactly the same way, and paper space limits generally will be set to emulate sheet sizes. Therefore, what you learn here will translate very easily when you begin using paper space.

> Begin a new drawing using "No Prototype".

After we are finished exploring the LIMITS command, we will create the new settings we want and save this drawing as your B-size prototype.

> Use F7 to turn the grid on.

> Type "Limits" or select "Drawing Limits" from the Data pull down menu.

AutoCAD will prompt you as follows:

```
Reset Model Space limits
ON/OFF/<\<>lower left corner> <\<>0.0000,0.0000>:
```

The ON and OFF options determine what happens when you attempt to draw outside the limits of the grid. With LIMITS off, nothing will happen. With LIMITS on, you will get a message that says "Attempt to draw outside of limits". Also, with LIMITS on, AutoCAD will not accept any attempt to begin an entity outside of limits, but will allow you to extend objects beyond the limits as long as they were started within them. By default, LIMITS is off, and we will leave it that way.

> Press enter.

This will enter the default values for the lower left corner (0,0). AutoCAD will give you a second prompt:

Upper right corner <12.0000,9.0000>:

Notice the default coordinates. These determine the size and shape of the grid you have been working with. We will set the limits to emulate a B-size sheet of paper.

> Type "18,12".

The grid will be regenerated to cover the screen, but you will not see any change in size. Under the new limits the complete grid has become larger than the present display, but you are only seeing part of it. Whenever you set limits larger or smaller than the current display, you will have to do a ZOOM All to see the display defined according to the new limits.

> Type "z" or select the Zoom All tool from the Standard toolbar.

> If necessary (i.e., if you are typing), type "a" or select "All".

You should have an 18 × 12 grid on your screen. Place the cursor on the upper right-hand grid point to check its coordinates. This is the grid you will use for a B-size prototype.

We suggest that you continue to experiment with setting limits, and that you try some of the possibilities listed in *Figure 4-1,* which is a table of sheet sizes.

> When you are done experimenting, return LIMITS to (0,0) and (18,12), using the LIMITS command, and then ZOOM All.

You are now in the drawing that we will use for your prototype, so it is not necessary to begin a new one for the next section.

TASK 2: Creating a Prototype

To make your own prototype, so that new drawings will begin with the settings you want, all you have to do is create a drawing that has those settings and then tell AutoCAD that this is the drawing you want to use to define your initial drawing setup. The first part should be easy for you now, since you have been doing your own drawing setup for each new drawing in this book.

> Make changes to the present drawing as follows:

GRID	1.00 ON (F7)	COORD	ON (F6)
SNAP	.25 ON (F9)	LTSCALE	.5
UNITS	2-place decimal	LIMITS	(0,0) (18,12)

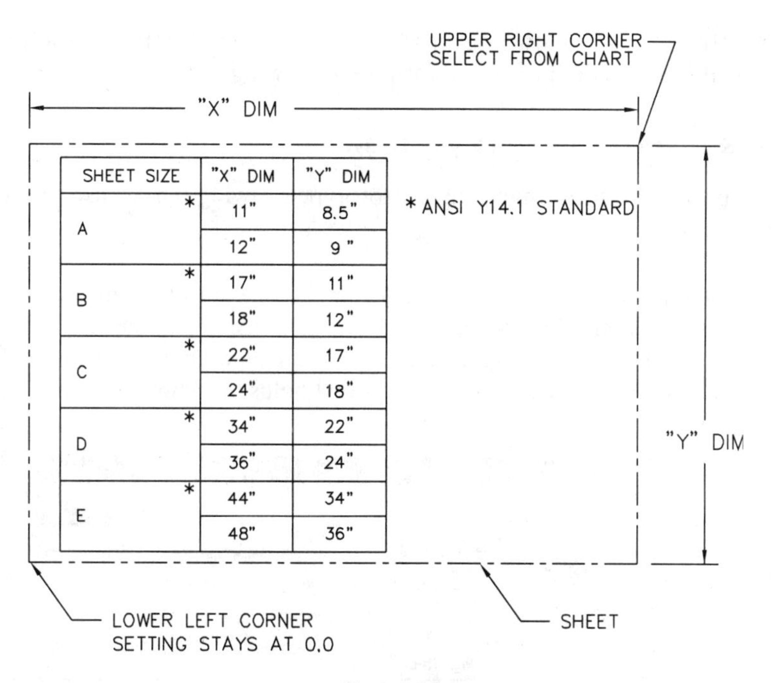

Figure 4-1

> Load all linetypes in the standard ACAD linetype file.
> Create the following layers and associated colors and linetypes.

Remember that you can make changes to your prototype at any time. The layers called "text", "hatch", and "dim" will not be used until Chapters 7 and 8, in which we introduce text, hatch patterns, and dimensions to your drawings. Creating them now will save time and avoid confusion later on.

Layer 0 is already defined.

LAYER NAME	STATE	COLOR	LINETYPE
0	On	7 (white)	CONTINUOUS
1	On	1 (red)	CONTINUOUS
2	On	2 (yellow)	HIDDEN
3	On	3 (green)	CENTER
TEXT	On	4 (cyan)	CONTINUOUS
HATCH	On	5 (blue)	CONTINUOUS
DIM	On	6 (magenta)	CONTINUOUS

> When all changes are made, save your drawing as "B".

NOTE: Do not leave anything drawn on your screen or it will come up as part of the prototype each time you open a new drawing. For some applications this may be useful. For now, we want a blank prototype.

If you have followed instructions up to this point, B.dwg should be on file. Now we will use it as the prototype for a new drawing.

TASK 3: Selecting a Prototype Drawing

The procedure for designating a prototype drawing makes use of a dialogue box and is quite simple.

> Type "new", select "New . . ." from the File pull down menu, or select the New tool at the far left of the Standard toolbar.

This will call up the familiar Create New Drawing dialogue box shown in *Figure 4-2.* The edit box on the top right holds the name of the current prototype draw-

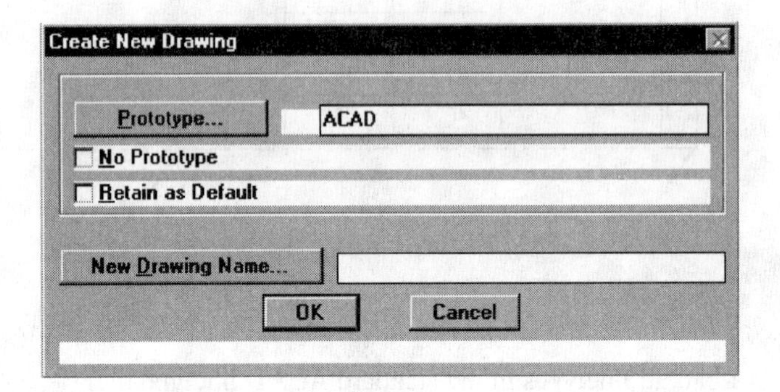

Figure 4-2

ing. In effect, the drawing named in this box will appear when a new drawing is created, unless you specifically tell AutoCAD to do otherwise. If there is no name in the box, then AutoCAD will use its own ACAD.dwg.

The simplest way to change the prototype is to type the name of a new prototype drawing in this box.

> To change the prototype drawing, double click inside the Prototype drawing name box.

If there is already a name in the box, it will become highlighted.

If the box is grayed out, you will notice that the box next to "No Prototype" is checked. Clicking in this box will remove the "x" and make the Prototype name box accessible.

If you prefer to select from a list, you can pick the "Prototype . . ." box, which will call up a standard file list box. You can select any file in the list to become the prototype for your new drawing.

> Type "b" or select B.dwg from the file list to make B.dwg the current prototype.

If your prototype is to be found in a directory other than the one in which ACAD.exe (the AutoCAD program) is found, you will need to include a drive designation. For example, "A:B" means B.dwg is on a floppy disk in drive A.

> Click on "OK".

Your B drawing is now designated as the prototype for the new drawing you are about to open. If you also want to retain B as the default prototype, so that it comes up every time you open any new drawing, you will need to click in the "Retain as Default" check box. If you are using this book in a course, ask your instructor about setting up a default prototype. Since other people probably use your workstation, your instructor will want to manage your use of a prototype carefully.

> Click inside the New Drawing Name edit box, and then type "4-1" or "A:4-1".
> Click on "OK" to exit the edit box and begin the new drawing.
You will find that the new drawing includes all the changes made to B.dwg.

TASK 4: Using the MOVE Command

The ability to copy and move objects on the screen is one of the great advantages of working on a CAD system. It can be said that CAD is to drafting as word processing is to typing. Nowhere is this analogy more appropriate than in the "cut and paste" capacities that the COPY and MOVE commands give you.

> Draw a circle with a radius of 1 somewhere near the center of the screen (9,6), as shown in *Figure 4-3*.

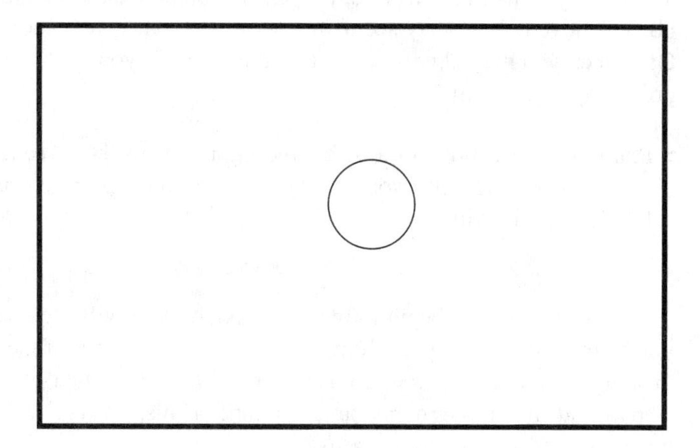

Figure 4-3

As discussed in Chapter 2, Release 13 allows you to pick objects before or after entering an edit command. In this exercise we will use MOVE both ways, beginning with the noun/verb or "pick first" method. We will pick the circle you have just drawn, but be aware that the selection set could include as many entities as you like and that a group of entities can be selected with a window or crossing box.
> Type "m" or select the Move tool from the Standard toolbar, as shown in *Figure 4-4*.
You will be prompted to select objects to move.
> Point to the circle.

Figure 4-4

As in the ERASE command, your circle will become dotted.

In the command area, AutoCAD will tell you how many objects have been selected and prompt you to select more. You may need to open the text screen (F2) to see this. When you are through selecting objects, you will need to press enter to move on.

> Press enter.

AutoCAD prompts:

Base point or displacement:

Most often you will show the movement by defining a vector that gives the distance and direction you want the object to be moved.

NOTE: In order to define movement with a vector, all AutoCAD needs is a distance and a direction. Therefore, the base point does not have to be on or near the object you are moving. Any point will do, as long as you can use it to show how you want your objects moved. This may seem strange at first, but it will soon become quite natural. Of course, you may choose a point on the object if you wish. With a circle, the center point may be convenient.

> Point to any location not too near the right edge of the screen.

AutoCAD will give you a rubber band from the point you have indicated and will ask for a second point:

Second point of displacement:

As soon as you begin to move the cursor, you will see that AutoCAD also gives you a circle to drag so you can immediately see the effect of the movement you are indicating. Let's say you want to move the circle 3.00 to the right. Watch the coordinate display and stretch the rubber band out until the display reads "3.00<0" (press F6 to get polar coordinates), as in *Figure 4-5*.

> Pick a point 3.00 to the right of your base point.

The rubber band and your original circle disappear, leaving you a circle in the new location.

Now, if ortho is on, turn it off (F8) and try a diagonal move. This time we will use the previous option to select the circle.

> Type "m" or select the Move tool, or press enter to repeat the command.

AutoCAD follows with the "Select objects:" prompt.

> Reselect the circle by typing "p" for previous.

> Press enter to end the object selection process.

> Select a base point.

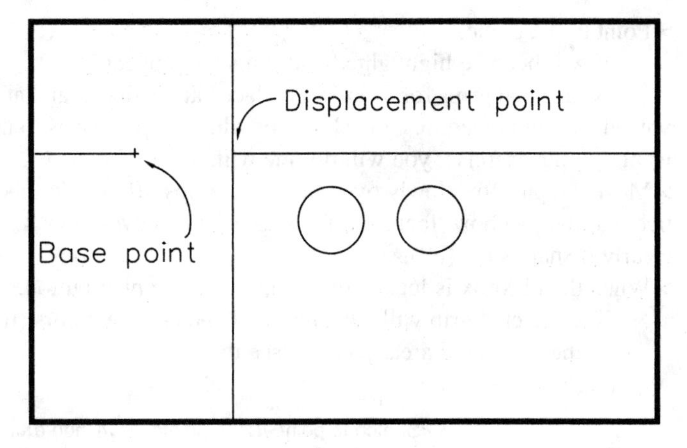

Figure 4-5

> Move the circle diagonally in any direction you like. *Figure 4-6* is an example of how this might look.

Try moving the circle back to the center of the screen. It may help to choose the center point of the circle as a base point this time, and choose a point at or near the center of the grid for your second point.

Moving with Grips

You can use grips to perform numerous editing procedures without ever entering a command. This is probably the simplest of all editing methods, called "autoediting". It does have some limitations, however. In particular, you can only select by pointing, windowing, or crossing.

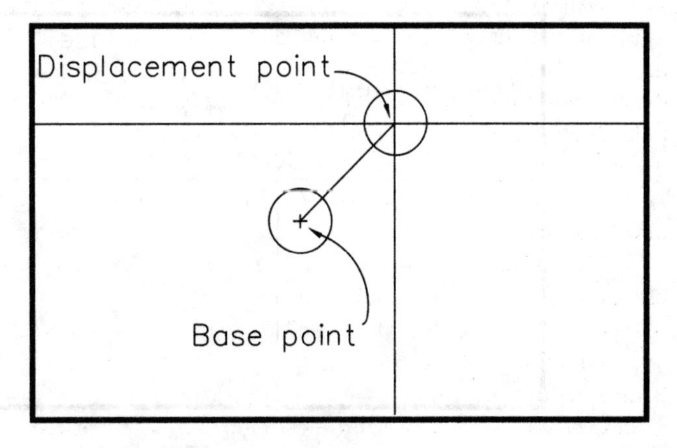

Figure 4-6

> Point to the circle.

It will become highlighted and grips will appear.

Notice that grips for a circle are placed at quadrants and at the center. In more involved editing procedures, the choice of which grip or grips to use for editing is significant. In this exercise, you will do fine with any of the grips.

> Move the pickbox slowly over one of the grips. If you do this carefully, you will notice that the pickbox "locks onto" the grip as it moves over it. You will see this more clearly if snap is off (F9).

> When the pickbox is locked on a grip, press the pick button.

The selected grip will become filled and change colors (from blue to red).

In the command area, you will see this:

```
** STRETCH **
<Stretch to point>/Base point/Copy/Undo/eXit:
```

Stretching is the first of a series of five autoediting modes that you can activate by selecting grips on objects. The word "stretch" has many meanings in AutoCAD and they are not always what you expect. For now, we will bypass stretch and use the MOVE mode.

> Press enter, the enter button on your cursor, or the space bar to bring up the MOVE autoedit mode.

You should see the following in the command area:

```
** MOVE **
<Move to point>/Base point/Copy/Undo/eXit:
```

Move the cursor now and you will see a rubber band from the selected grip to the same position on a dragged circle, as illustrated in *Figure 4-7*.

> Pick a point anywhere on the screen.

The circle will move where you have pointed.

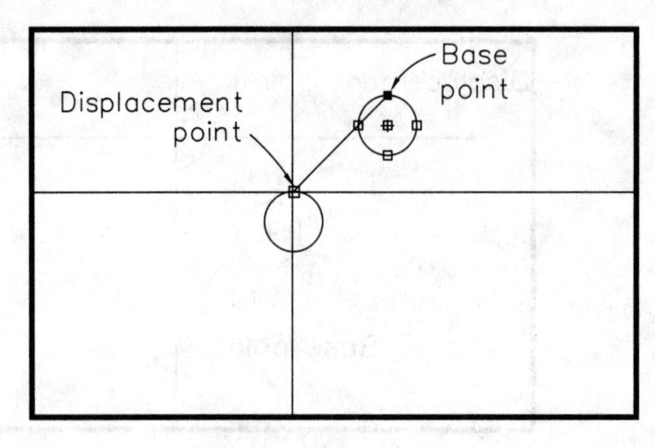

Figure 4-7

TASK 5: Using the COPY Command

The COPY command works so much like the MOVE command that you should find it quite easy to learn at this point. The main difference is that the original object will not disappear when the second point of the displacement vector is given. Also there is an additional option, to make multiple copies of the same object, which we will explore in a moment.

First, we suggest that you try making several copies of the circle in various positions on the screen. Try using both noun/verb and verb/noun sequences. Notice that "c" is not an alias for COPY. Also, notice that there is a Copy tool on the Standard toolbar that initiates the COPYCLIP command. This tool has a very different function. It is used to copy objects to the Windows Clipboard and then into other applications. It has no effect on objects within your drawing, other than to save them on the clipboard.

When you are satisfied that you know how to use the basic COPY command, move on to the MULTIPLE copy option.

The Multiple Copy Option

What this option does is allow you to show a whole series of vectors starting at the same base point, and AutoCAD will place copies of your selection set accordingly.

> Type "Copy" or select the Copy tool from the Modify toolbar, as shown in *Figure 4-8.*

Figure 4-8

> Point to one of the circles on your screen.
> Press enter to end the selection process.
> Type "m".
> Show AutoCAD a base point.
> Show AutoCAD a second point.

You will see a new copy of the circle, and notice also that the prompt for a "Second point of displacement" has returned in the command area. AutoCAD is waiting for another vector, using the same base point as before.

> Show AutoCAD another second point.
> Show AutoCAD another second point.

Repeat as many times as you wish. If you get into this, you may begin to feel like a magician pulling ring after ring out of thin air and scattering them across the screen. The results will appear something like *Figure 4-9.*

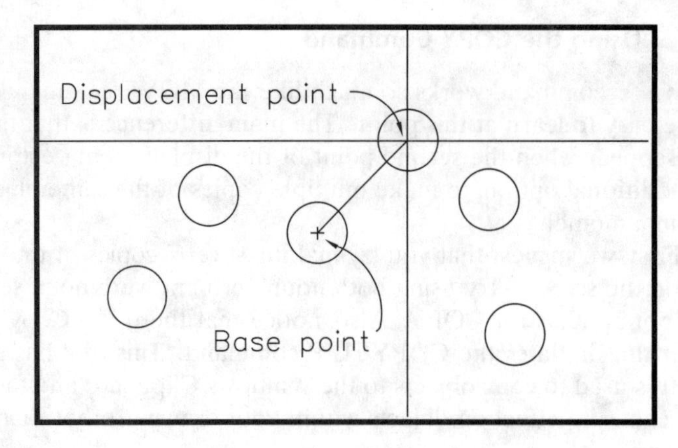

Figure 4-9

Copying with Grips

The grip editing system includes a variety of special techniques for creating multiple copies in all five modes. For now we will stick with the copy option in the MOVE mode, which provides a shortcut method for creating the same kind of process you just went through with the COPY command.

Since you should have several circles on your screen now, we will take the opportunity to demonstrate how you can use grips on more than one object at a time.

> Pick any two circles.

The circles you pick should become highlighted and grips should appear on both, as illustrated in *Figure 4-10*.

> Pick any grip on either of the two highlighted circles.

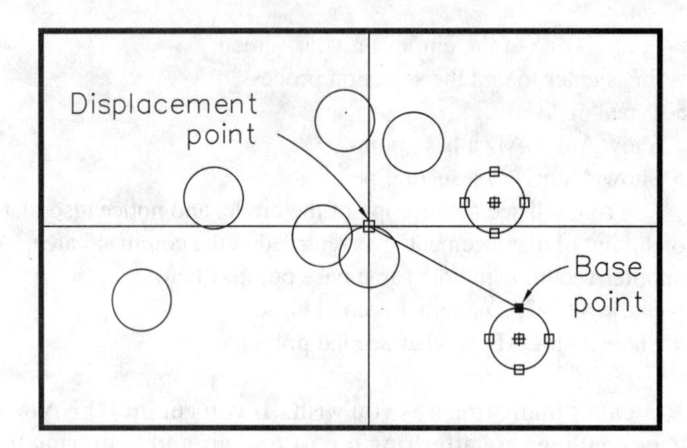

Figure 4-10

The grip should change colors.

> Press enter, the enter button on your cursor, or the space bar.

This should bring you into MOVE mode. Notice the prompt:

```
** MOVE **
<Move to point>/Base point/Copy/Undo/eXit:
```

> Type "c" to initiate copying.

The prompt will change to:

```
** MOVE (multiple) **
<Move to point>/Base point/Copy/Undo/eXit:
```

You will find that all copying in the grip editing system is multiple copying. Once in this mode, AutoCAD will continue to create copies wherever you press the pick button until you exit by typing "x" or pressing enter.

> Move the cursor and observe the two dragged circles.
> Pick a point to create copies of the two highlighted circles.
> Pick another point to create two more copies.
> When you are through, press enter to exit the grip editing system.
> Press Esc twice to remove grips.

TASK 6: Using the ARRAY Command—Rectangular Arrays

The ARRAY command gives you a powerful alternative to simple copying. It takes an object or group of objects and copies it a specific number of times in mathematically defined, evenly spaced, locations. An array is a repetition in matrix form of the same figure.

There are two types of arrays. Rectangular arrays are linear and defined by rows and columns. Polar arrays are angular and based on the repetition of objects around the circumference of an arc or circle. The dots on the grid are an example of a rectangular array; the lines on any circular dial are an example of a polar array.

Both types are common. We will explore rectangular arrays in this chapter and polar arrays in the next.

In preparation for this exercise, erase all the circles from your screen. This is a good opportunity to try the ERASE All option.

> Type "e" or select the Erase tool from the Modify toolbar.
> Type "all".
> Press enter.
> Now draw a single circle, radius .5, centered at the point (2,2).
> Type "Array" or select the Rectangular Array tool from the Copy flyout on the Modify toolbar, as shown in *Figure 4-11.*

You will see the "Select objects" prompt.

> Point to the circle.

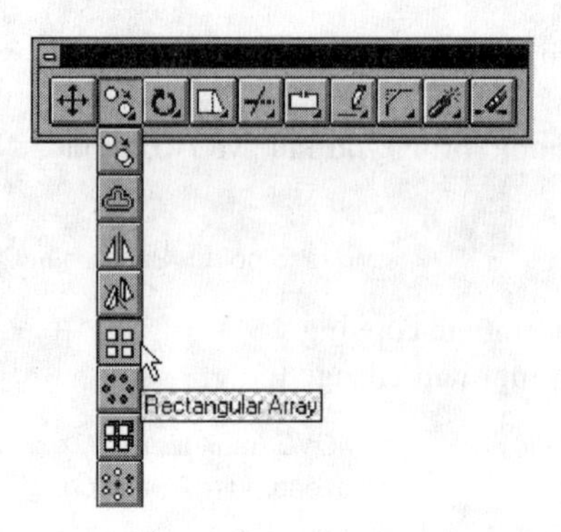

Figure 4-11

> Press enter to end the selection process.

AutoCAD will ask which type of array you want:

Rectangular or Polar array (R/P) <R>:

> Type "r". You can also press enter if R is the default.

AutoCAD will prompt you for the number of rows in the array.

Number of rows (---) <1>:

The (---) is to remind you of what a row looks like, i.e., it is horizontal.

The default is 1, so if you press enter you will get a single row of circles. The number of circles in the row will depend, then, on the number of columns you specify. We will ask for three rows instead of just one.

> Type "3".

AutoCAD now asks for the number of columns in the array:

Number of columns (|||) <1>:

Using the same format (|||), AutoCAD reminds you that columns are vertical. The default is 1 again. What would an array with three rows and only one column look like?

We will construct a five-column array.

> Type "5".

Now AutoCAD needs to know how far apart to place all these circles. There will be 15 of them in this example—three rows with five circles in each row. AutoCAD prompts:

Unit cell or distance between rows (---):

"Unit cell" means that you can respond by showing two corners of a window. The horizontal side of this window would give the space between columns; the vertical side would give the space between rows. You could do this exercise by showing a 1 ×

1 window. We will use the more basic method of typing values for these distances. The distance between rows will be a vertical measure.

> Type "1".
AutoCAD now asks for the horizontal distance between columns:

Distance between columns (|||):

> Type "1" again.
You should have a 3 × 5 array of circles, as shown in *Figure 4-12.*

Notice that AutoCAD builds arrays up and to the right. This is consistent with the coordinate system, which puts positive values to the right on the horizontal x axis and upwards on the vertical y axis. Negative values can be used to create arrays in other directions.

TASK 7: Changing Plot Configuration Parameters

In the previous chapter, you began using the plot preview feature of the Plot Configuration Dialogue box. In this chapter, we encourage you to continue to learn about plot configuration by exploring several more areas of the dialogue box. We have used no specific drawing for illustration. Now that you know how to use plot preview, you can observe the effects of changing plot parameters with any drawing you like and decide at any point whether you actually want to print out the results. We will remind you continually to look at a plot preview after making any change in configuration. This is the best way to learn about plotting.

We will start with device selection and move on to pen parameters and additional parameters. In the next chapter, we will complete the introduction to the dialogue box with paper size, orientation, scaling, rotation, and origin.

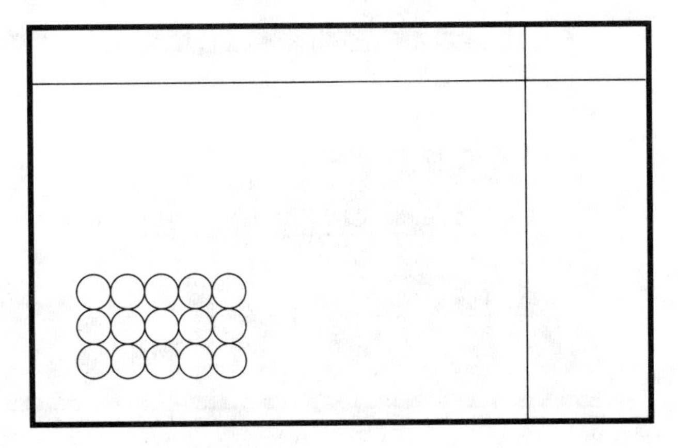

Figure 4-12

> To begin this exploration, you should have a drawing or drawn objects on your screen so that you can observe the effects of various configuration changes you will be making.
> Type "plot", select "Print . . ." from the "File" pull down menu, or select the Print tool from the Standard toolbar.

This will open up the Plot Configuration dialogue box.

Device and Default Information

The device box should contain the name of the printer or plotter you are planning to use. If not, you can get a list of available options through the subdialogue illustrated in *Figure 4-13.*

> Click on "Device and Default Selection . . ." or type "d".

You should see a subdialogue box similar to the one shown in the figure. The list of devices will be your own, of course. If there is a need to change plotters or printers, you can do so by selecting from the list. If the device you want is not on the list, you will need to make sure the device driver is properly installed and then use the CONFIG command to add it to the list. See the *AutoCAD Installation Guide For Windows* for additional information.

Along with the list of available plotting devices, this dialogue box allows you to save plotting specifications to ASCII files and get them back later. AutoCAD automatically saves your most recent plot parameters in a file called ACAD.cfg. However, for a more permanent file, you can create a separate file. Such files are saved with a "pcp" extension ("Plot Configuration Parameters"). There are many parameters to

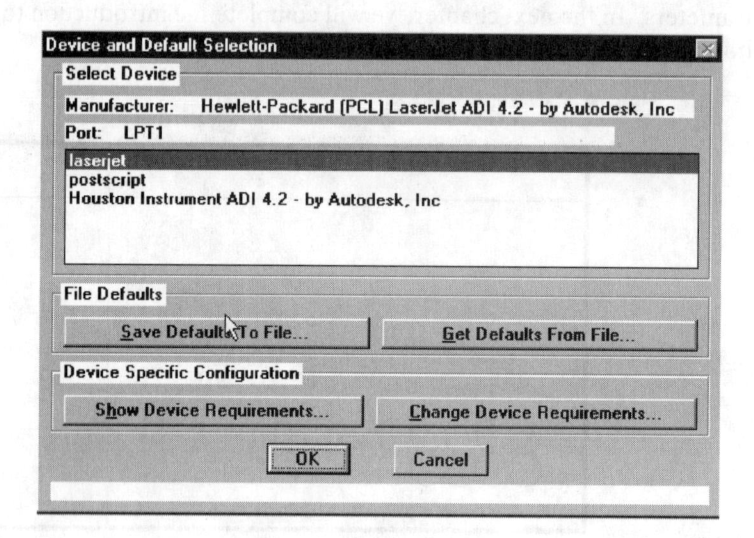

Figure 4-13

deal with in plotting, including as many as 256 pen colors. Therefore, saving a set of parameters to a file may be far more efficient than specifying them manually each time you plot. You may have one set of parameters for one device, another set for a different device, or you may have different setups for different types of drawings.

In addition, some plotting devices have special configuration requirements not needed or not available on other devices. If this is the case with your device, the boxes under "Device Specific Configuration" will be accessible. The specific parameters and options will be determined by the device driver for your plotter. See the *AutoCAD Installation Guide For Windows* for more information.

> Click on "OK" or "Cancel" to exit the Device and Default Selection dialogue.

This will bring you back to the Plot Configuration dialogue box. Assuming the device named in the Device and Default Information box is the correct one, you are ready to move on. However, if you have made any changes, we recommend that you do a plot preview. Changing devices may bring about changes in paper size and orientation, for example, and you will see these changes reflected graphically in the partial and full previews.

Pen Parameters

If you are using a plotter with multiple pens, you can assign colors, linetypes, widths, and speeds to each pen individually. If you are using a printer, you will find that the Pen Assignments box is grayed out. If you do have access, we suggest you look at the subdialogue now, even though you may not want to make changes.

> Click on "Pen Assignments . . ." or type "p".

This will call up the dialogue box shown in *Figure 4-14*. If you select any pen, you will see its color, number, linetype, speed, and width displayed in the Modify Values box at the right. Changes can be made in these edit boxes in the usual manner.

Pen Assignments

Color	Pen No.	Linetype	Speed	Pen Width
1	1	0	4	0.010
2	2	0	4	0.010
3	3	0	4	0.010
4	4	0	4	0.010
5	5	0	4	0.010
6	6	0	4	0.010
7	7	0	4	0.010
8	8	0	4	0.010
9	1	0	24	0.010
10	2	0	24	0.010

Modify Values
Color:
Pen:
Ltype:
Speed:
Width:

Feature Legend...
Pen Width:
OK
Cancel

Figure 4-14

The specifications made here are designed to relate pen numbers on your plotter to pen colors. If pen numbers are correctly related to pen colors, then layers will automatically be plotted in their assigned colors. The linetypes assigned to layers in AutoCAD are also plotted automatically and do not need to be assigned to pens at this point.

NOTE: If your plotter supports multiple linetypes, you will reach a subdialogue showing numbered linetypes by clicking on "Feature Legend . . .". These are not to be confused with the linetypes created within your AutoCAD drawing associated with layers. They should be used in special applications to vary the look of "continuous" AutoCAD lines only; otherwise, you will get a confused mixture of linetypes when your plotter tries to break up lines that AutoCAD has already drawn broken.
> Click on "OK" or "Cancel" to exit the "Pen Assignments" subdialogue.

Additional Parameters

This is a crucial area of the dialogue box that allows you to specify the portion of your drawing to be plotted. You have some familiarity with this from Chapter 2 when you plotted using a window selection. Changes here will have a very significant impact on the effective plotting area. Be sure to use plot preview any time you make changes in these parameters.

Also available in this area are hidden line removal (for 3D drawings); fill area adjustment, which affects the way pen width is interpreted when drawing wide lines; and plotting to a file instead of to an actual plotter.

The radio buttons on the left show the options for plotting area. "Display" will create a plot using whatever is actually on the screen. If you have used the ZOOM command to enlarge a portion of the drawing before entering PLOT and then select this option, AutoCAD will plot whatever you have zoomed in on. "Extents" refers to the actual drawing area in which you have drawn objects. It may be larger or smaller than limits of the drawing. Drawing limits, as you know, are specified by you using the LIMITS command. If you are using our standard "B" prototype, they will go from (0,0) to (18,12). "View" will be accessible only after you have defined views in the drawing using the VIEW command. "Window" will not be accessible until you define a window (if you have defined a window in a previous drawing, it may be saved in ACAD.cfg, the plot configuration file).

> As an exercise, we suggest that you try switching among Display, Extents, Limits, and Window selections and use plot preview to see the results. Use both partial and full previews to ensure that you can clearly see what is happening. This can be very instructive.

Whenever you make a change, also observe the changes in the boxes showing Plotted Inches = Drawing Units. Assuming that "Scaled to Fit" is checked, you will see significant changes in these scale ratios as AutoCAD adjusts scales according to how much area it is being asked to plot.

We will explore the last two areas of Plot Configuration, "Paper Size and Orientation" and "Scale, Rotation, and Origin", in the next chapter. In the meantime, you should continue to experiment with plotting and plot previewing as you complete the drawings that follow.

TASKS 8, 9, and 10

All the drawings in this chapter will use your new prototype. The settings and layers should be as you have defined them. Do not expect, however, that you will never need to change them. Layers will stay the same throughout this book, but limits will change from time to time and grid and snap will change frequently.

The main thing you should be focused on in doing these drawings is to become increasingly familiar with the COPY and ARRAY commands. When you finish each drawing, go into Plot Configuration and try a plot preview even if you do not intend to print it out on paper.

DRAWING 4-1: GRILL

This drawing should go very quickly if you use the ARRAY command.

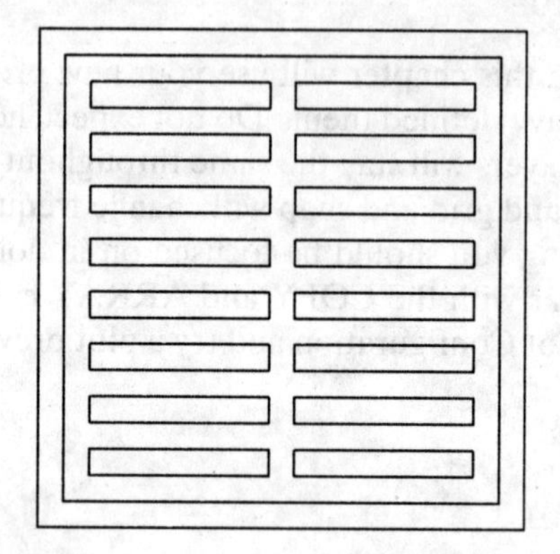

DRAWING SUGGESTIONS

GRID = .5
SNAP = .25

> Begin with a 4.75 × 4.75 square.
> Move in .25 all around to create the inside square.
> Draw the rectangle in the lower left-hand corner first, then use the ARRAY command to create the rest.
> Also remember that you can undo a misplaced array using the U command.

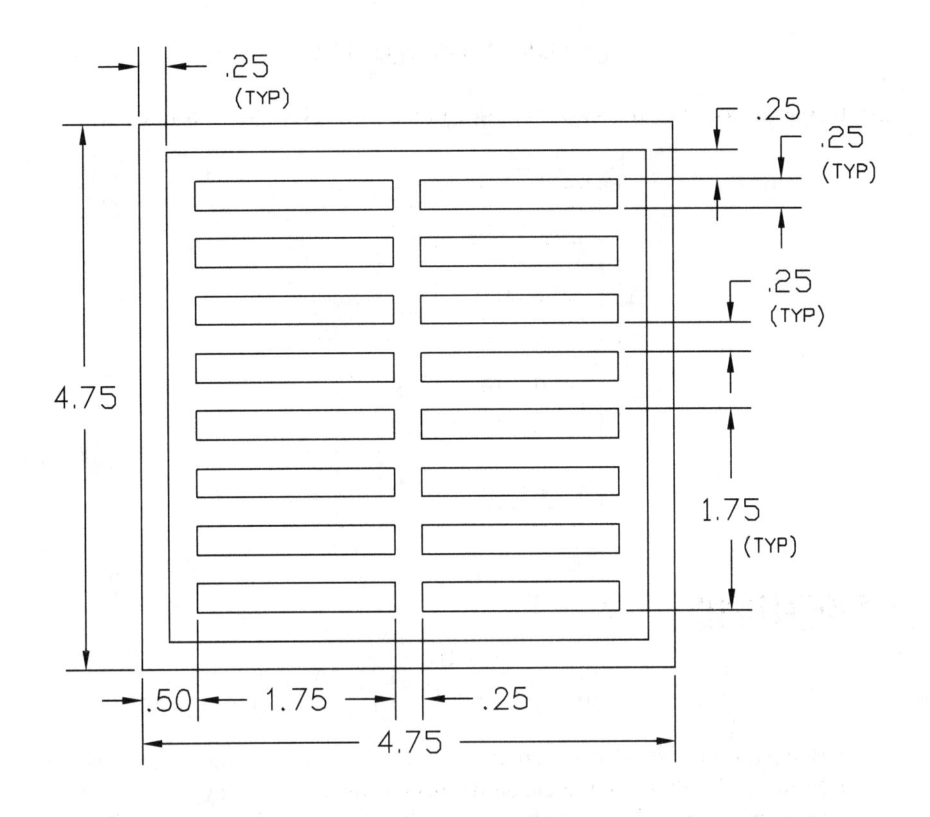

GRILL
Drawing 4–1

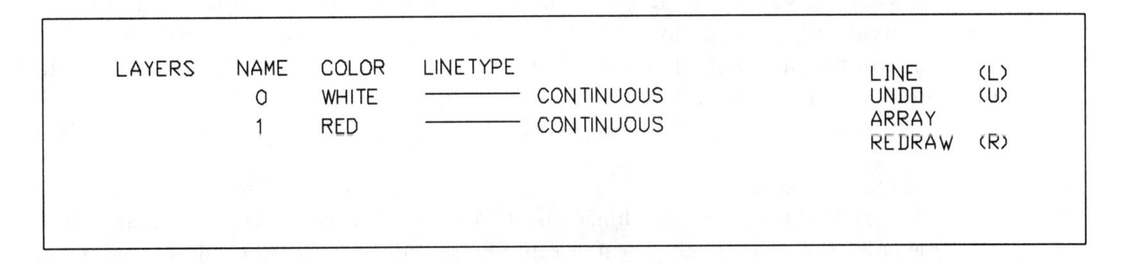

F1 HELP
F2 TEXT/GRAPHICS SCREEN
F6 ABSOLUTE/OFF/POLAR COORDS
F7 ON/OFF GRID
F8 ON/OFF ORTHO
F9 ON/OFF SNAP

DRAWING 4-2: TEST BRACKET

This is a great drawing for practicing much of what you have learned up to this point. Notice the suggested snap, grid, ltscale, and limit settings, and use the ARRAY command to draw the 25 circles on the front view.

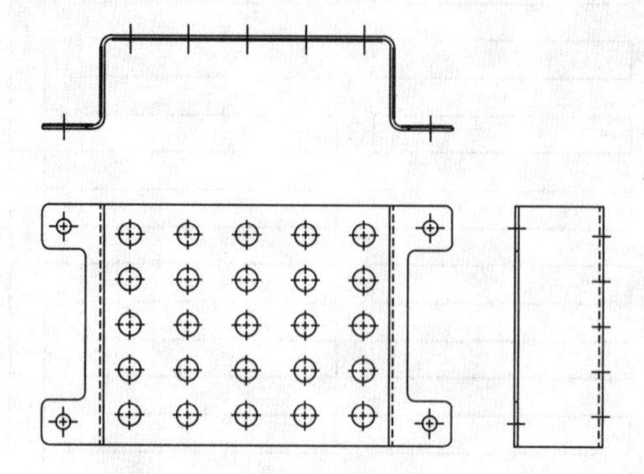

DRAWING SUGGESTIONS

GRID = .5	LTSCALE = .50
SNAP = .25	LIMITS = (0,0) (24,18)

> Be careful to draw all lines on the correct layers, according to their linetypes.
> Draw center lines through the circles before copying or arraying them; otherwise, you will have to go back and draw them on each individual circle or repeat the array process.
> A multiple copy will work nicely for the four .50 diameter holes. A rectangular array is definitely desirable for the twenty-five .75 diameter holes.

CREATING CENTER MARKS WITH THE DIMCEN SYSTEM VARIABLE

There is a simple way to create the center marks and center lines shown on all the circles in this drawing. It involves changing the value of a dimension variable called "dimcen" (dimension center). Dimensioning and dimension variables are discussed in Chapter 8, but if you would like to jump ahead, the following procedure will work nicely in this drawing.

> Type "dimcen".

The default setting for dimcen is .09, which will cause AutoCAD to draw a simple cross as a center mark. Changing it to −.09 will tell AutoCAD to draw a cross that reaches across the circle.

> Type "−.09".
> After drawing your first circle, and before arraying it, type "dim". This will put you in the dimension command.
> Type "cen", indicating that you want to draw a center mark. This is a very simple dimension feature.
> Point to the circle.

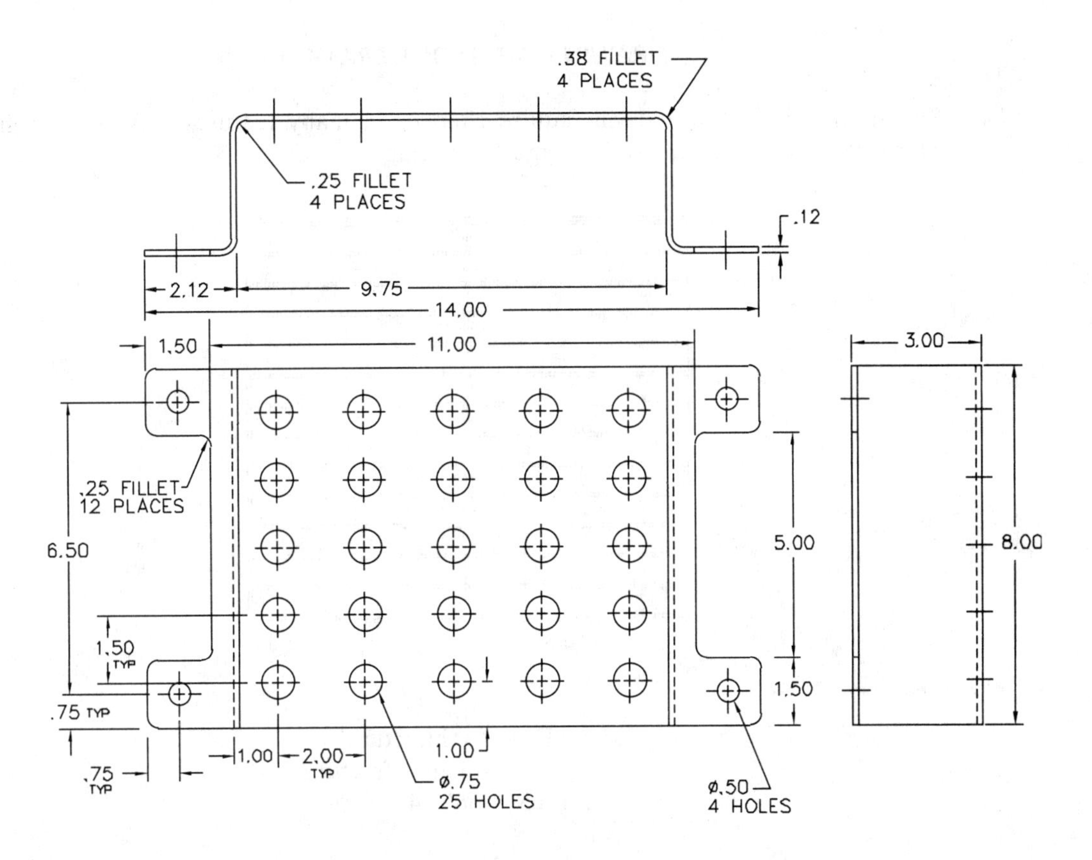

TEST BRACKET

Drawing 4–2

LAYERS	NAME	COLOR	LINETYPE	
	0	WHITE	————	CONTINUOUS
	1	RED	————	CONTINUOUS
	2	YELLOW	- - - - - - -	HIDDEN
	3	GREEN	—— - ——	CENTER

LINE (L)
CIRCLE (C)
FILLET
ARRAY
ZOOM (Z)

F1	F2	F6	F7	F8	F9
HELP	TEXT/GRAPHICS SCREEN	ABSOLUTE/OFF/POLAR COORDS	ON/OFF GRID	ON/OFF ORTHO	ON/OFF SNAP

DRAWING 4-3: FLOOR FRAMING

This architectural drawing will require changes in many features of your drawing setup. Pay close attention to the suggested settings.

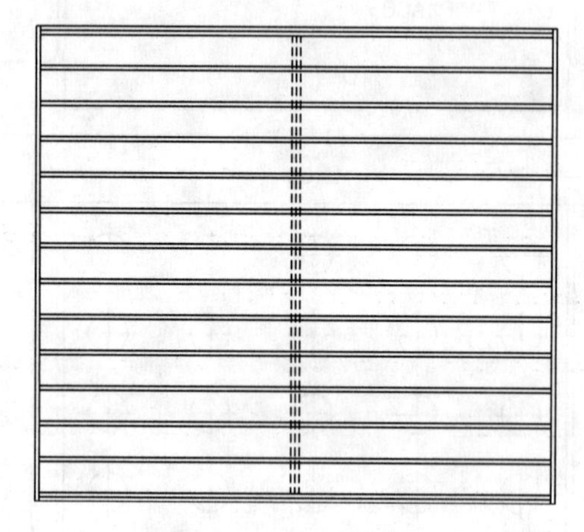

DRAWING SUGGESTIONS

UNITS = Architectural,
smallest fraction = 1″
LIMITS = 36′, 24′
GRID = 1′
SNAP = 2″
LTSCALE = 12

> Be sure to use foot (′) and inch (″) symbols when setting limits, grid, and snap (but not ltscale).

> Begin by drawing the 20′ × 17′10″ rectangle, with the lower left corner somewhere in the neighborhood of (4′,4′).

> Complete the left and right 2 × 10 joists by copying the vertical 17′10″ lines 2″ in from each side. You may find it helpful to use the arrow keys when working with such small increments.

> Draw a 19′8″ horizontal line 2″ up from the bottom and copy it 2″ higher to complete the double joists.

> Array the inner 2 × 10 in a 14-row by 1-column array, with 16″ between rows.

> Set to layer 2 and draw the three 17′4″ hidden lines down the center.

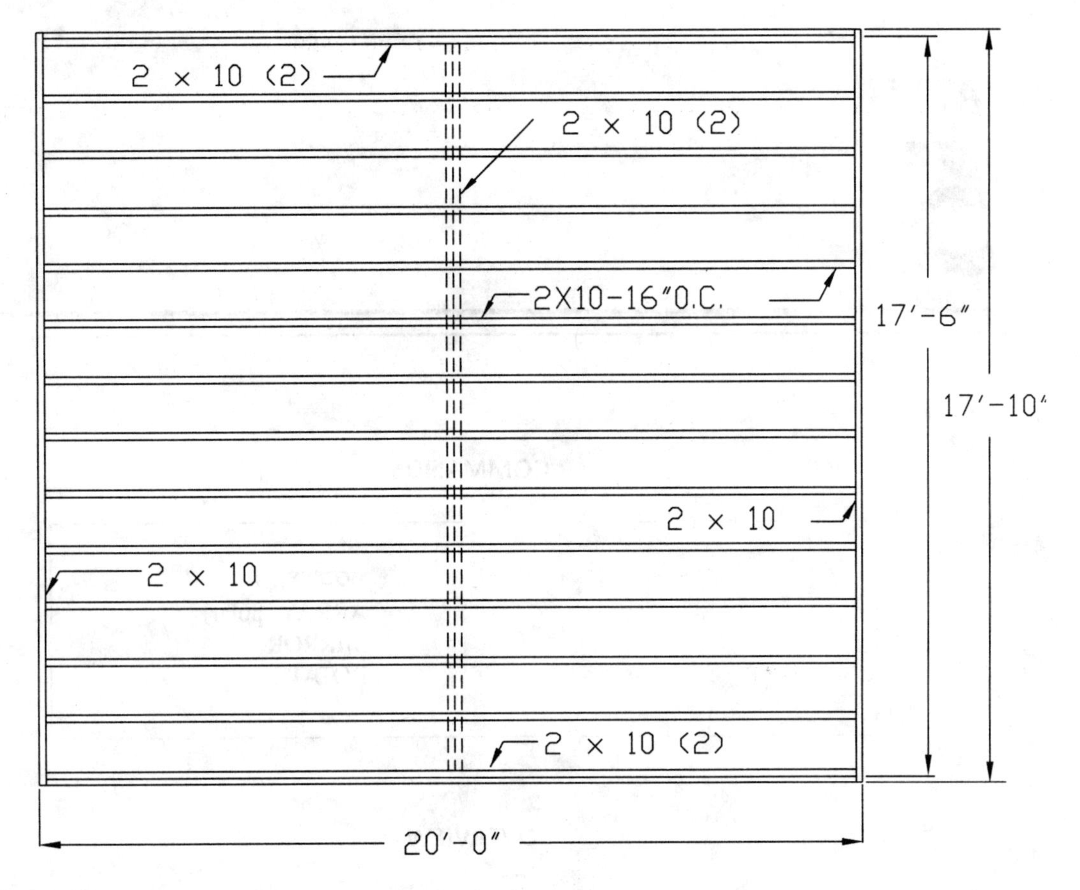

FLOOR FRAMING

Drawing 4-3

LAYERS	NAME	COLOR	LINETYPE	
	0	WHITE	————	CONTINUOUS
	1	RED	————	CONTINUOUS
	2	YELLOW	- - - - -	HIDDEN

LINE (L)
COPY
ARRAY
REDRAW (R)
ZOOM (Z)

F1	F2	F6	F7	F8	F9
HELP	TEXT/GRAPHICS SCREEN	ABSOLUTE/OFF/POLAR COORDS	ON/OFF GRID	ON/OFF ORTHO	ON/OFF SNAP

CHAPTER 5

COMMANDS

DRAW	MODIFY
ARC	ARRAY (polar)
	MIRROR
	ROTATE

OVERVIEW

So far, every drawing you have done has been composed of lines and circles. In this chapter you will learn a third major entity, the ARC. In addition, you will expand your ability to manipulate objects on the screen. You will learn to ROTATE objects and create their MIRROR images. First, we will pick up where we left off in Chapter 4 by showing you how to create polar arrays.

TASKS

1. Create three polar arrays.
2. Draw arcs in eight different ways.
3. Rotate a previously drawn object.
4. Create mirror images of previously drawn objects.
5. Change Paper Size, Orientation, Plot Scale, Rotation, and Origin.
6. Do Drawing 5-1 ("Dials").
7. Do Drawing 5-2 ("Alignment Wheel").
8. Do Drawing 5-3 ("Hearth").

TASK 1: Creating Polar Arrays

The procedure for creating polar arrays is lengthy and requires some explanation. The first two steps are the same as in rectangular arrays. Step 3 is also the same, except that you respond with "polar" or "p" instead of "rectangular" or "r". From here on, the steps will be new. First you will pick a center point, and then you will have several options for defining the array.

There are three qualities that define a polar array, but two are sufficient. A polar array is defined by two of the following: 1) a certain number of items, 2) an angle that these items span, and 3) an angle between each item and the next. However you define your polar array, you will have to tell AutoCAD whether or not to rotate the newly created objects as they are copied.

> Begin a new drawing using the B prototype.
> In preparation for this exercise, draw a vertical 1.00 line at the bottom center of the screen, near (9.00,2.00) as shown in *Figure 5-1.* We will use a 360 degree polar array to create *Figure 5-2.*
> Type "Array" or select the Array tool from the Copy flyout on the Modify toolbar.
> Select the line.
> Type "p" or select "polar".

So far, so good. Nothing new up to this point. Now you have a prompt that looks like this:

Center point of array:

Rectangular arrays are not determined by a center, so we did not encounter this prompt before. Polar arrays, however, are built by copying objects around the circumferences of circles or arcs, so we need a center to define one of these.

> Pick a point directly above the line and somewhat below the center of the screen. Something in the neighborhood of (9.00,4.50) will do. The next prompt is:

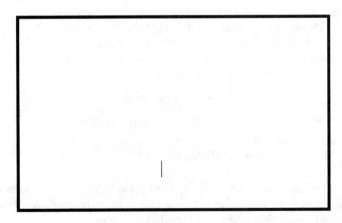

Figure 5-1

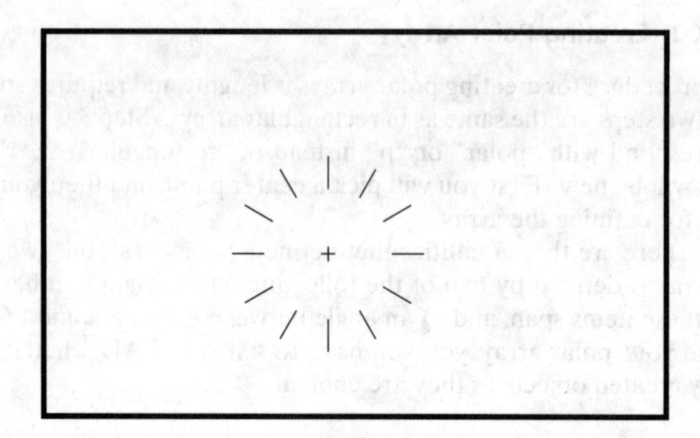

Figure 5-2

Number of items:

Remember that we have a choice of two out of three among number of items, angle to fill, and angle between items. This time we will give AutoCAD the first two.
> Type "12".

Now that AutoCAD knows that we want 12 items, all it needs is either the angle to fill with these or the angle between the items. It will ask first for the angle to fill:

Angle to fill (+=ccw,-=cw) <360>:

The symbols in parentheses tell us that if we give a positive angle the array will be constructed counterclockwise; if we give a negative angle, it will be constructed clockwise. Get used to this; it will come up frequently.

The default is 360 degrees, meaning an array that fills a complete circle.

If we did not give AutoCAD an angle (that is, if we responded with a "0"), we would be prompted for the angle between. We will give 360 as the angle to fill.

> Press enter to accept the default, a complete circle.

AutoCAD now has everything it needs, except that it doesn't know whether we want our lines to retain their vertical orientation or to be rotated along with the angular displacement as they are copied. AutoCAD asks:

Rotate objects as they are copied? <Y>:

Notice the default, which we will accept.
> Press enter or type "y".
Your screen should resemble *Figure 5-2*.

This ends our discussion of polar arrays. With the options AutoCAD gives you, there are many possibilities that you may want to try out. As always, we encourage experimentation. When you are satisfied, erase everything on the screen and do a REDRAW in preparation for learning the ARC command.

TASK 2: Drawing Arcs

Learning AutoCAD's ARC command is an exercise in geometry. In this section, we will give you a firm foundation for understanding and drawing arcs so that you will not be confused by all the options that are available. The information we give you will be more than enough to do the drawings in this chapter. Refer to the AutoCAD Command Reference and the chart at the end of this section (*Figure 5-4*) if you need additional information.

AutoCAD gives you eight distinct ways to draw arcs, and if you count variations in order, 11. Every option requires you to specify three pieces of information: where to begin the arc, where to end it, and what circle it is theoretically a part of.

We will begin by drawing an arc using the simplest method, which is also the default: the 3-point option. The geometric key to this method is that any three points not on the same line determine a circle or an arc of a circle. AutoCAD uses this in the CIRCLE command (the 3P option) as well as in the ARC command.

> Type "a" or select the Arc tool from the Draw toolbar, as in *Figure 5-3.*
AutoCAD's response will be this prompt:

Center/<Start point>:

Accepting the default by specifying a point will leave you open to all those options in which the start point is specified first.

If you type a "C", AutoCAD will prompt for a center point and follow with those options that begin with a center.

> Select a starting point near the center of the screen.
AutoCAD prompts:

Center/End/<Second point>:

We will continue to follow the default, three-point sequence by specifying a second point. You may want to refer to the chart (*Figure 5-4*) as you draw this arc.

> Select any point one or two units away from the previous point. Exact coordinates are not important.

Once AutoCAD has two points, it gives you an arc to drag. By moving the cursor slowly in a circle and in and out, you can see the range of what the third point will produce.

AutoCAD also knows now that you have to provide an end point to complete the arc, so the prompt has only one option:

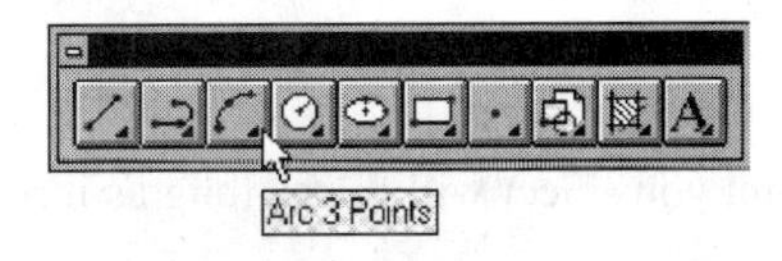

Figure 5-3

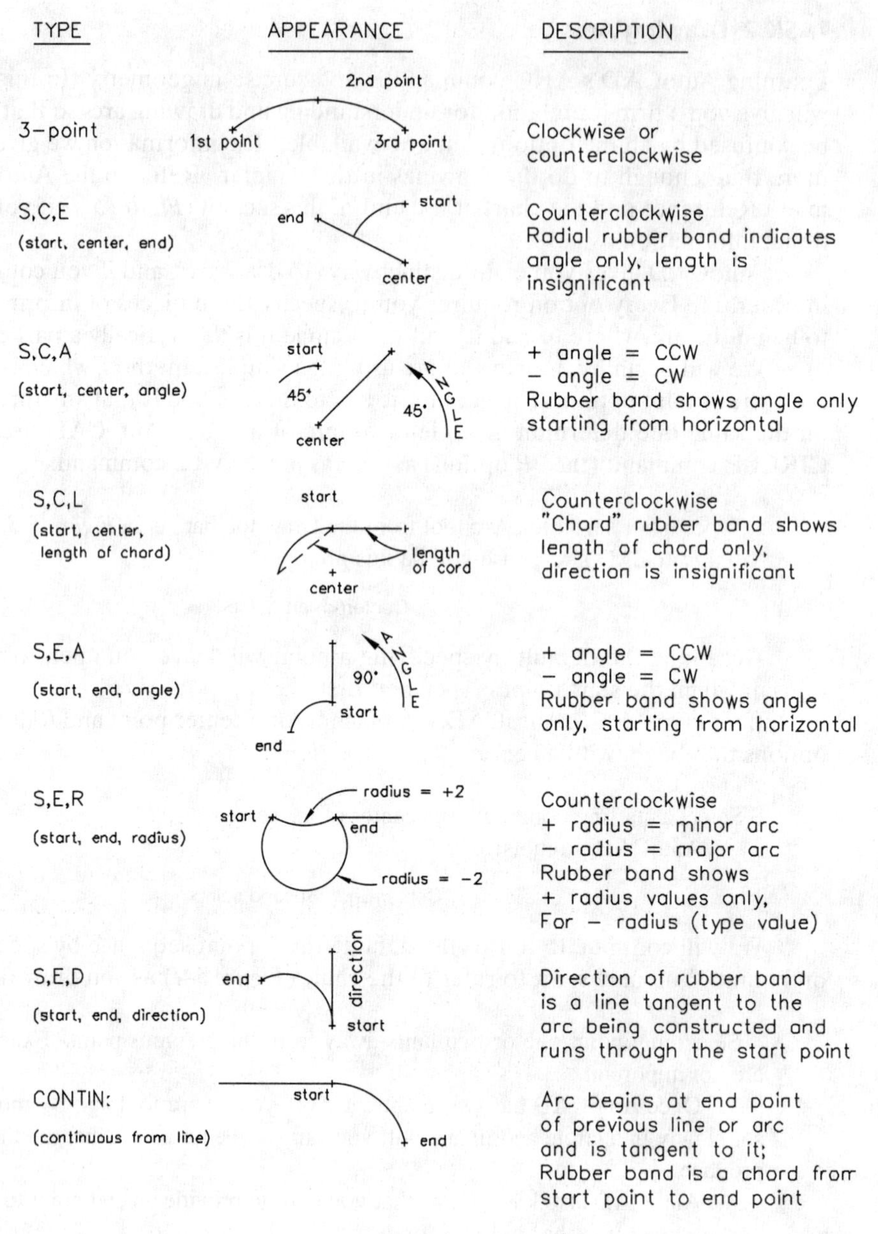

TYPE
APPEARANCE
DESCRIPTION
3-point
2nd point
1st point
3rd point
Clockwise or counterclockwise
S,C,E
(start, center, end)
start
end
center
Counterclockwise
Radial rubber band indicates angle only, length is insignificant
S,C,A
(start, center, angle)
start
45°
center
ANGLE
+ angle = CCW
− angle = CW
Rubber band shows angle only starting from horizontal
S,C,L
(start, center, length of chord)
start
length of cord
center
Counterclockwise
"Chord" rubber band shows length of chord only, direction is insignificant
S,E,A
(start, end, angle)
90°
start
end
ANGLE
+ angle = CCW
− angle = CW
Rubber band shows angle only, starting from horizontal
S,E,R
(start, end, radius)
radius = +2
start
end
radius = −2
Counterclockwise
+ radius = minor arc
− radius = major arc
Rubber band shows
+ radius values only,
For − radius (type value)
S,E,D
(start, end, direction)
end
direction
start
Direction of rubber band is a line tangent to the arc being constructed and runs through the start point
CONTIN:
(continuous from line)
start
end
Arc begins at end point of previous line or arc and is tangent to it;
Rubber band is a chord from start point to end point

Figure 5-4

End point:

Any point you select will do, as long as it produces an arc that fits on the screen.
> Pick an end point.

As you can see, three-point arcs are easy to draw. It is much like drawing a line, except that you have to specify three points instead of two. In practice, however, you do not always have three points to use this way. This necessitates the broad range of options in the ARC command. The dimensions you are given and the objects already drawn will determine what options are useful to you.

Next, we will create an arc using the start,center, end method, the second option illustrated in *Figure 5-4.*

> Type "u" to undo the three-point arc.
> Type "a" or select the Arc tool.
> Select a point near the center of the screen as a start point.

The prompt that follows is the same as for the three-point option, but we will not use the default this time:

Center/End/<Second point>:

We will choose the Center option.

> Type "c".

This tells AutoCAD that we want to specify a center point next, so we see this prompt:

Center:

> Select any point roughly one to three units away from the start point.

The circle from which the arc is to be cut is now clearly determined. All that is left is to specify how much of the circle to take, which can be done in one of three ways, as the prompt indicates:

Angle/Length of chord/<End point>:

We will simply specify an end point by typing coordinates or pointing. First, move the cursor slowly in a circle and in and out to see how the method works. As before, there is an arc to drag, and now there is a radial direction rubber band as well. If you pick a point anywhere along this rubber band, AutoCAD will assume you want the point where it crosses the circumference of the circle.

NOTE: Here, as in the polar arrays in this chapter, AutoCAD is building arcs counterclockwise, consistent with its coordinate system.

> Select an end point to complete the arc.

Now that you have tried two of the basic methods for constructing an arc, we strongly suggest that you study the chart and then try out the other options. The notes in the right-hand column will serve as a guide to what to look for.

The differences in the use of the rubber band from one option to the next can be confusing. You should understand, for instance, that in some cases the linear rubber band is only significant as a distance indicator; its angle is of no importance and is ignored by AutoCAD. In other cases, it is just the reverse. The length of the rubber band is irrelevant, while its angle of rotation is important.

NOTE: One additional trick you should try out as you experiment with arcs is as follows: If you press enter or the space bar at the "Center/<Start point>" prompt, AutoCAD will use the end point of the last line or arc you drew as the new starting point and construct an arc tangent to it. This is the same as the Continue option on the pull down menu.

This completes the present discussion of the ARC command. Constructing arcs, as you may have realized, can be tricky. Another option that is available and often useful is to draw a complete circle and then use the TRIM or BREAK commands to cut out the arc you want. BREAK and TRIM are introduced in the next chapter.

TASK 3: Using the ROTATE Command

ROTATE is a fairly straightforward command, and it has some uses that might not be apparent immediately. For example, it frequently is easier to draw an object in a horizontal or vertical position first and then ROTATE it than it would be to draw it diagonally.

In addition to the ROTATE command there is also a rotate mode in the grip edit system, which we will introduce at the end of the exercise.

> In preparation for this exercise, clear your screen and draw a horizontally oriented arc near the center of your screen, as in *Figure 5-5.* Exact coordinates and locations are not important.

We will begin by rotating the arc to the position shown in *Figure 5-6.*

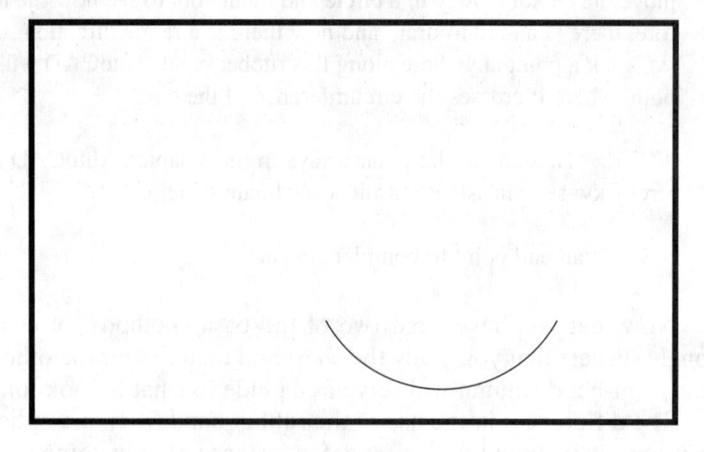

Figure 5-5

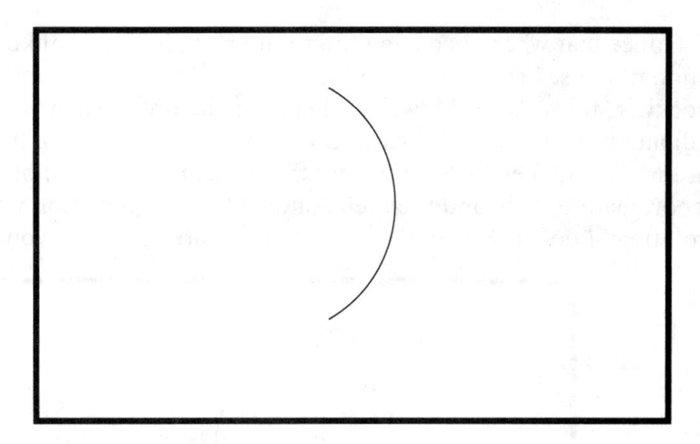

Figure 5-6

> Select the arc.
> Type "Rotate" or select the Rotate tool from the Modify toolbar, as shown in *Figure 5-7.*

You will be prompted for a base point.

Base point:

This will be the point around which the object is rotated. The results of the rotation, therefore, are dramatically affected by your choice of base point. We will choose a point at the left tip of the arc.
> Point to the left tip of the arc.

The prompt that follows looks like this:

<Rotation angle>/Reference:

The default method is to indicate a rotation angle directly. The object will be rotated through the angle specified and the original object deleted.

Move the cursor in a circle and you will see that you have a copy of the object to drag into place visually. If ortho or snap are on, turn them off to see the complete range of rotation.
> Type "90" or point to a rotation of 90 degrees (use F6 if your coordinate display is not showing polar coordinates).

The results should resemble *Figure 5-6.*

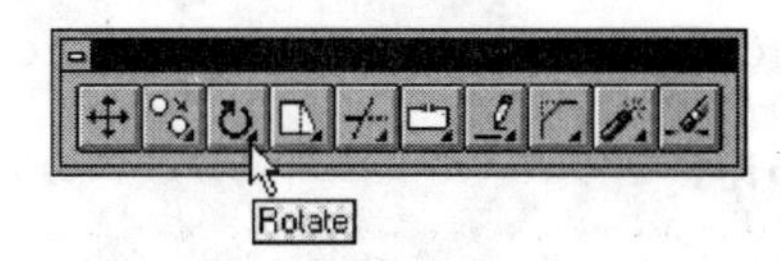

Figure 5-7

Notice that when specifying the rotation angle directly like this, the original orientation of the selected object is taken to be 0 degrees. The rotation is figured counterclockwise from there. However, there may be times when you want to refer to the coordinate system in specifying rotation. This is the purpose of the "reference" option. To use it, all you need to do is specify the present orientation of the object relative to the coordinate system, and then tell AutoCAD the orientation you want it to have after rotation. Look at *Figure 5-8.* To rotate the arc as shown, you either can indicate a

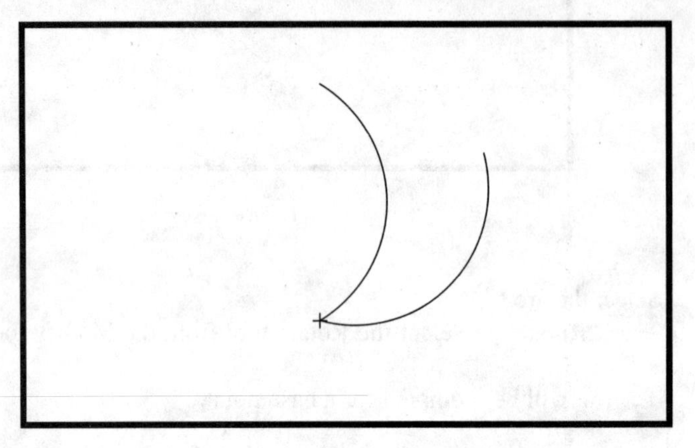

Figure 5-8

rotation of -45 degrees or tell AutoCAD that it is presently oriented to 90 degrees and you want it rotated to 45 degrees. Try this method for practice.

> Press enter to repeat the ROTATE command.
> Select the arc.
> Press enter to end selection.
> Choose a base point at the lower tip of the arc.
> Type "r" or select "Reference".

AutoCAD will prompt for a reference angle:

Reference angle <0>:

> Type "90".

AutoCAD prompts for an angle of rotation:

Rotation angle:

> Type "45".

Your arc should now resemble the solid arc in *Figure 5-8.*

Rotating with Grips

Rotating with grips is simple and there is a very useful option for copying, but your choice of object selection methods is limited, as always, to pointing and windowing. Try this:

> Pick the arc.

The arc will become highlighted and grips will appear.

> Pick the center grip.

> Press enter twice to reach the Rotate mode.

Move your cursor in a circle and you will see the arc rotating around the grip at the center of the arc.

> Now, type “b”.

This will allow you to pick a base point other than the selected grip.

> Pick a base point above and to the left of the grip, as shown in *Figure 5-9.*

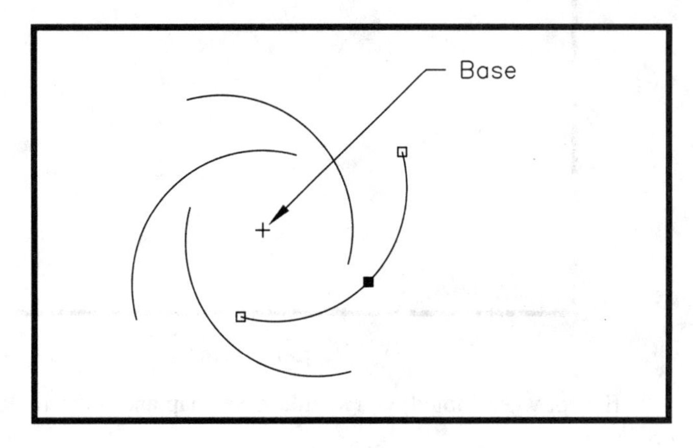

Figure 5-9

Move your cursor in circles again. You will see the arc rotating around the new base point.

> Type “c”.

Notice the Command area prompt, which indicates you are now in a rotate and multiple copy mode.

> Pick a point showing a rotation angle of 90 degrees, as illustrated by the top arc in *Figure 5-9.*

> Pick a second point showing a rotation angle of 180 degrees, as illustrated by the arc at the left in the figure.

> Pick point 3 at 270 degrees to complete the design shown in *Figure 5-9.*

> Press enter to exit the grip mode.

This capacity to create rotated copies is very useful, as you will find when you do the drawings at the end of the chapter.

TASK 4: Creating MIRROR Images of Objects on the Screen

There are two main differences between the command procedures for MIRROR and ROTATE. First, in order to mirror an object, you will have to define a mirror line; second, you will have an opportunity to indicate whether you want to retain the original object or delete it. In the ROTATE sequence, the original is always deleted.

There is also a mirror mode in the grip edit system, which we will explore at the end of the task.

> To begin this exercise, undo the rotate copy process so that you are left with a single arc. Rotate it and move it to the left so that you have a bowl-shaped arc placed to the left of the center of your screen, as in *Figure 5-10*.

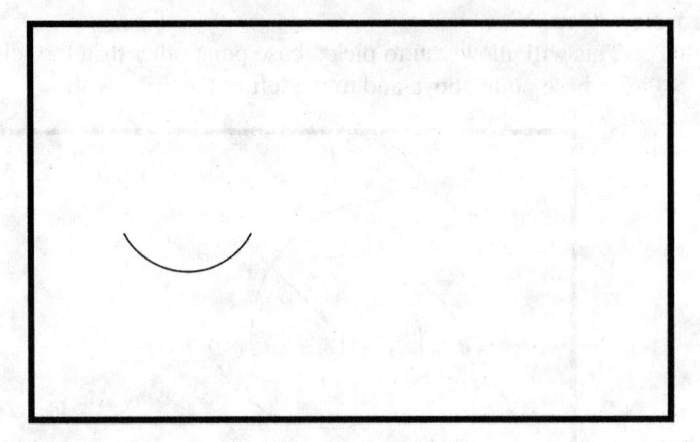

Figure 5-10

Except where noted, you should have snap and ortho on to do this exercise.

> Select the arc.

> Type "Mirror" or select the Mirror tool from the Copy flyout on the Modify toolbar, as shown in *Figure 5-11*.

Now AutoCAD will ask you for the first point of a mirror line.

First point of mirror line:

A mirror line is just what you would expect; all points on your object will be mirrored across the line at a distance equal and opposite to their distance away from it.

We will show a mirror line even with the top of the arc, so that the end points of the mirror images will be touching.

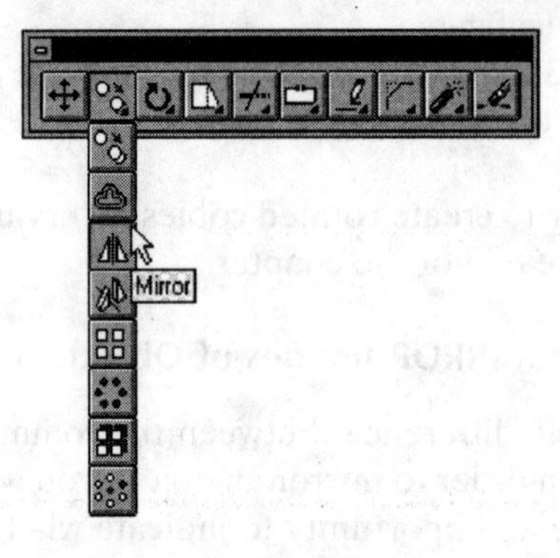

Figure 5-11

> Select a point even with the left end point of the arc, as in *Figure 5-12*.

You are prompted to show the other end point of the mirror line:

Second point:

The length of the mirror line is not important. All that matters is its orientation. Move the cursor slowly in a circle, and you will see an inverted copy of the arc moving with you to show the different mirror images that are possible, given the first point you have specified. Turn ortho off to see the whole range of possibilities, then turn it on again to complete the exercise.

We will select a point at 0 degrees from the first point, so that the mirror image will be directly above the original arc and touching at the end points as in *Figure 5-12*.

> Select a point directly to the right (0 degrees) of the first point.

The dragged object will disappear until you answer the next prompt, which asks if you want to delete the original object or not.

Delete old objects? <N>:

This time around, we will not delete the original.

> Press enter to retain the old object. Your screen will look like *Figure 5-12,* without the mirror line in the middle.

Mirroring with Grips

Mirror is the fifth grip mode, after stretch, move, rotate, and scale. It works exactly like the rotate mode, except that the rubber band will show you a mirror line instead of a rotation angle. The option to retain or delete the original is obtained through the copy option, just as in the rotate mode. Try it.

> Select the two arcs on your screen by pointing or windowing.

The arcs will be highlighted and grips will be showing.

> Pick any of the grips.

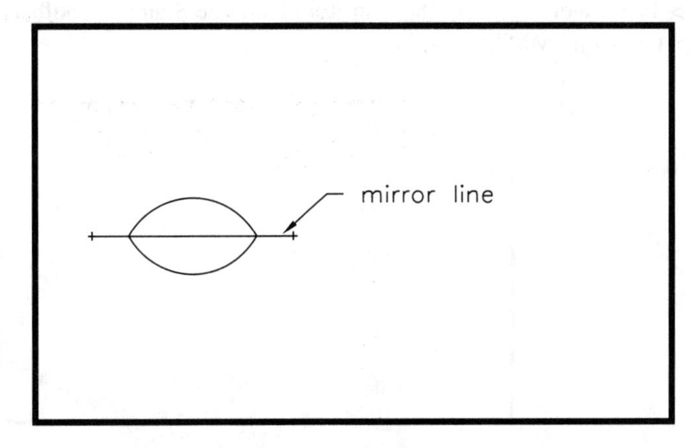

Figure 5-12

> Pass to the MIRROR mode by pressing enter four times, or by typing "mi".

Move the cursor and observe the dragged mirror images of the arcs. Notice that the rubber band operates as a mirror line, just as in the MIRROR command.

> Type "b".

This frees you from the selected grip and allows you to create a mirror line from any point on the screen. Notice the "Base point:" prompt in the command area.

> Type "c" or select "Copy".

As in the rotate mode, this is how you retain the original in a grip edit mirroring sequence.

> Pick a base point above and to the right of the arcs.

> Pick a second point directly below the first.

Your screen should resemble *Figure 5-13.*

> Press enter to end the mirror/copy sequence.

TASK 5: Changing Paper Size, Orientation, Plot Scale, Rotation, and Origin

This discussion will complete our exploration of the basic features of the Plot Configuration dialogue box. The parameters involved will dramatically impact the results of your hardcopy output. In particular, putting all of these variables together with each other and previously discussed parameters takes a lot of skill and know how. You can gain this expertise through experience with actual plotting and plot previewing. As you become more comfortable with what is going on, it is especially valuable to plot using more than one device and different paper sizes.

Paper Size and Orientation

This box gives critical information and choices about the size and orientation of your drawing sheet. To begin with, you can choose to see information presented in inches or millimeters by selecting one of the two radio buttons. Inches is the default.

> Type "plot" or select the Print tool from the Standard toolbar.

> Click on "MM".

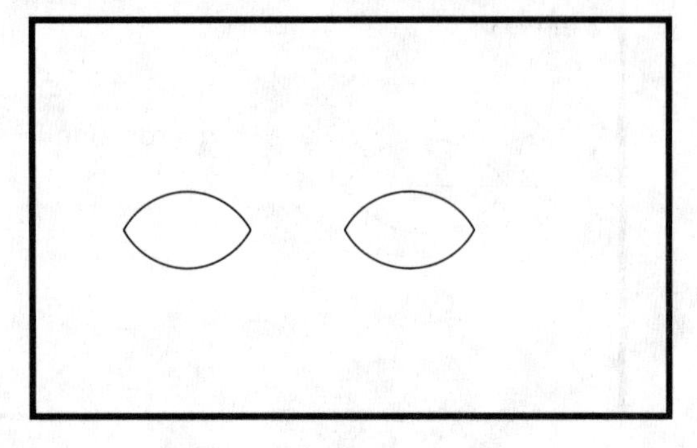

Figure 5-13

You will see that both the plot area and the scale values are changed to reflect metric units.

> Click on "Inches".

On the right, you will see a rectangle showing the orientation of the paper in the plotting device. It will be either landscape (horizontal) or portrait (vertical). This orientation is part of the device driver and cannot be changed directly. If you want to plot horizontally when your device is configured vertically, or vice-versa, you must rotate the plot using the Rotation and Origin subdialogue described in the next section.

The plot area is shown below the radio buttons and is also determined by the plotting device. The default size will be the maximum size available on your printer or plotter. If you are using a printer, for example, you may have the equivalent of an A-size sheet. In this case, the plot area will be close to 8.00 × 11.00.

> Click on "Size . . .".

This will open the Paper Size subdialogue box shown in *Figure 5-14*. The standard sizes listed in the box on the left are device specific and will depend on your plotter. If you are using a printer, for example, you may see only the A size on the list. The "USER" boxes on the right allow you to define plotting areas of your own. This means, for example, that you can plot in a 5.00 × 5.00 area on an 8.00 × 11.00 or larger sheet of paper.

Paper size can be changed by picking any of the sizes on the list at the left.

> Change the paper size, if you wish, and then click on "OK" to exit the subdialogue.

If you have changed sizes, the new size will be named next to the Size box, and the new plot area will be listed below. The change also may be reflected in the scale box ("Plotted Inches = Drawing Units").

Once again, if you have changed paper sizes we suggest that you do a plot preview. Paper size is one of the important factors in determining effective plotting area.

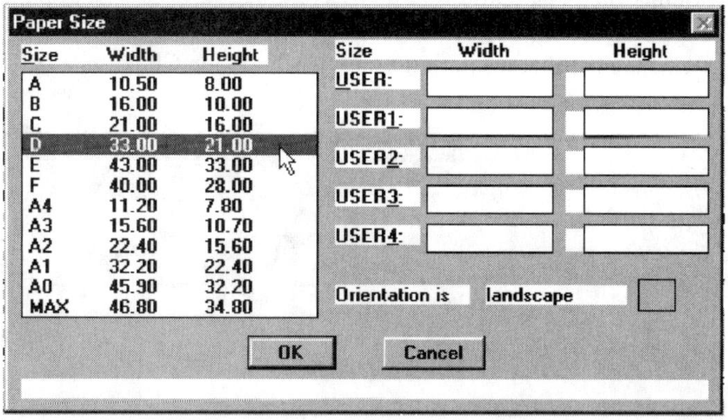

Paper Size

Size	Width	Height
A	10.50	8.00
B	16.00	10.00
C	21.00	16.00
D	33.00	21.00
E	43.00	33.00
F	40.00	28.00
A4	11.20	7.80
A3	15.60	10.70
A2	22.40	15.60
A1	32.20	22.40
A0	45.90	32.20
MAX	46.80	34.80

Size Width Height
USER:
USER1:
USER2:
USER3:
USER4:
Orientation is landscape
OK Cancel

Figure 5-14

Scale, Rotation, and Origin

Plots can be scaled to fit the available plot area, or given an explicit scale of paper units to drawing units. "Scaled to Fit" is the default, as shown by the check box. With this setting, the area chosen for plotting in the Additional Parameters box (Display, Extents, Limits, etc.) will be plotted as large as possible within the plot area specified in the Paper Size and Orientation box. Fitting the chosen area to the available plot area will dictate the scale shown in the edit boxes. Plotted inches or millimeters are shown in relation to drawing units. If either the area to be plotted or the paper size is changed, the change will be reflected in the scale boxes.

If "Scaled to Fit" is not checked, the scale will default to 1 = 1. In this case, drawing units will be considered equivalent to paper size units. 1 to 1 scale is definitely preferred when plotting from Paper Space (Chapter 6). Other scales can be specified explicitly by typing in the edit boxes. When you change the drawing scale, the area to be plotted may be affected. If the area is too large for the paper size given the scale, then only a portion of the chosen area will be plotted. If the area becomes smaller than the available paper, some blank space will be left. Changing scales should be followed by a partial preview.

The area of the paper that is actually used for plotting may also be affected by the rotation and origin of the plot. If your device is configured with paper oriented vertically (portrait style), and you want to plot horizontally (landscape), then you will need to rotate 90 degrees. This is done with the radio buttons in the Plot Rotation and Origin subdialogue box.

> Click in the Rotation and Origin box.

This will call up the dialogue box shown in *Figure 5-15*. You can rotate to 0, 90, 180, or 270 degrees as shown. Rotating a plot by 90 degrees will have a very significant impact on the relationship between paper size and effective drawing area.

This box will also allow you to change the origin of the plot. Plots usually originate in the lower left-hand corner of the page, but you can alter this. For example, let's say you wanted to plot a 5 × 5 area in the upper right of an 8 × 11 sheet, landscape orientation. You could do this by creating a 5 × 5 "User" paper size and then moving the origin from (0,0) to (6,3).

> Change the plot rotation if you wish, and then Click on "OK" to exit the dialogue box.

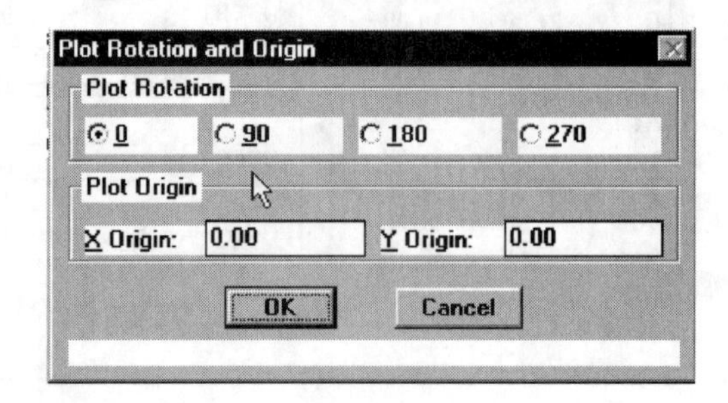

Figure 5-15

At this point, you have been through all the major parameters of plot configuration. It is important that you gain experience in using the configuration parameters available. When you have a drawing ready, open the dialogue box and make whatever adjustments you think are necessary. You should access at least one partial and one full preview along the way. When everything looks right, check your plotting device and paper and then click on "OK" to start it rolling.

TASKS 6, 7, 8, 9, and 10

You have learned some complex sequences in this chapter, especially in the ARC and polar ARRAY commands, so take your time doing these drawings and be sure that you understand the commands involved. Your knowledge of AutoCAD and CAD operation is increasing rapidly at this point, and it will be important that you practice what you have learned carefully.

DRAWING 5-1: DIALS

This is a relatively simple drawing that will give you some good practice with polar arrays and the ROTATE and COPY commands.

Notice that the needle drawn at the top of the next page is only for reference; the actual drawing includes only the plate and the three dials with their needles.

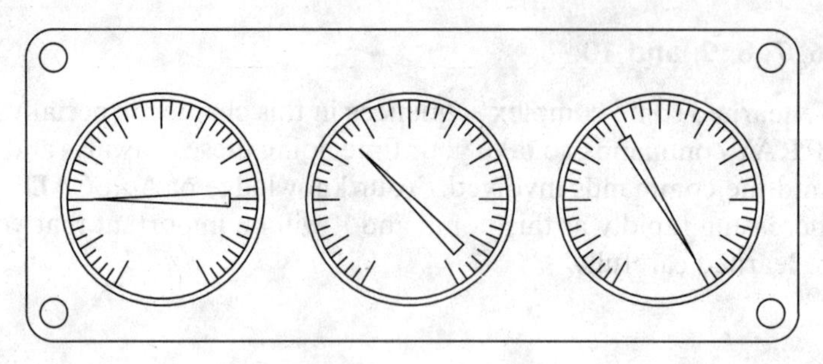

DRAWING SUGGESTIONS

GRID = .25	LTSCALE = .50
SNAP = .125	LIMITS = (0,0) (12,9)

> After drawing the outer rectangle and screw holes, draw the left-most dial, including the needle. Draw a .50 vertical line at the top and array it to the left (counterclockwise-a positive angle) and to the right (negative) to create the 11 larger lines on the dial. Use the same operation to create the 40 small (.25) markings.

> Complete the first dial and then use a multiple copy to produce two more dials at the center and right of your screen. Be sure to use a window to select the entire dial.

> Finally, use the ROTATE command to rotate the needles as indicated on the new dials. Use a window to select the needle, and rotate it around the center of the dial.

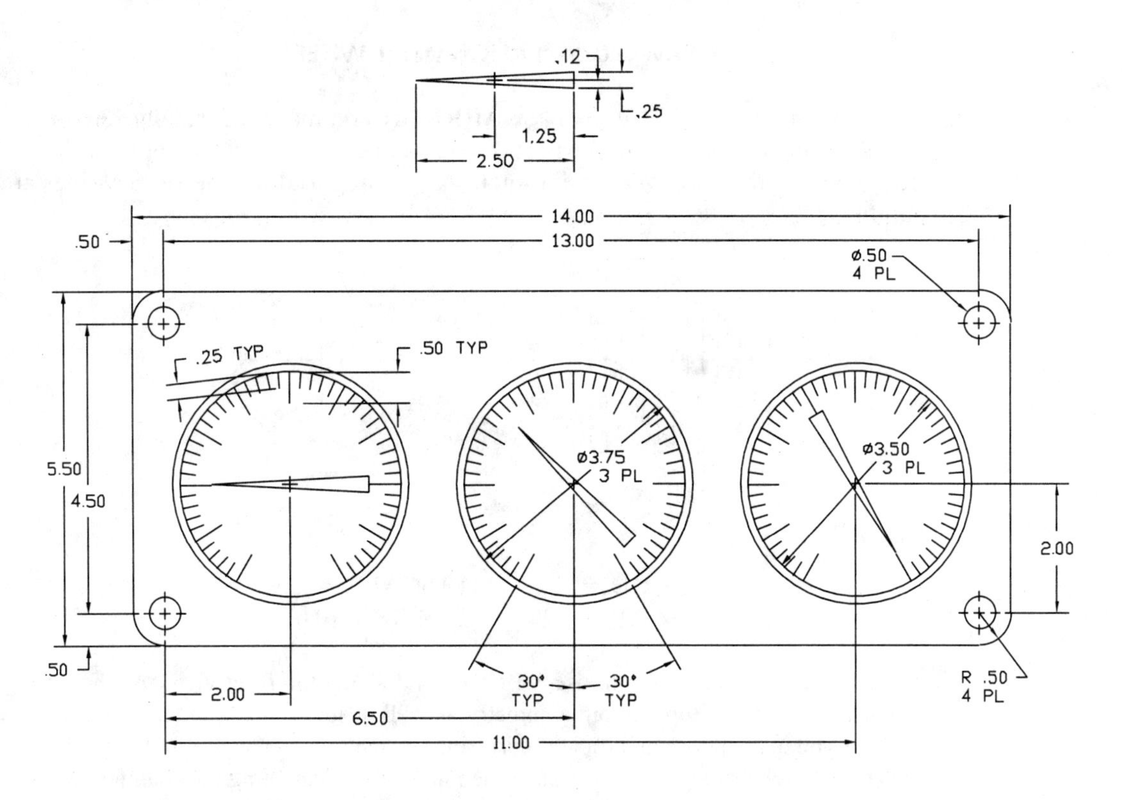

DIALS

Drawing 5–1

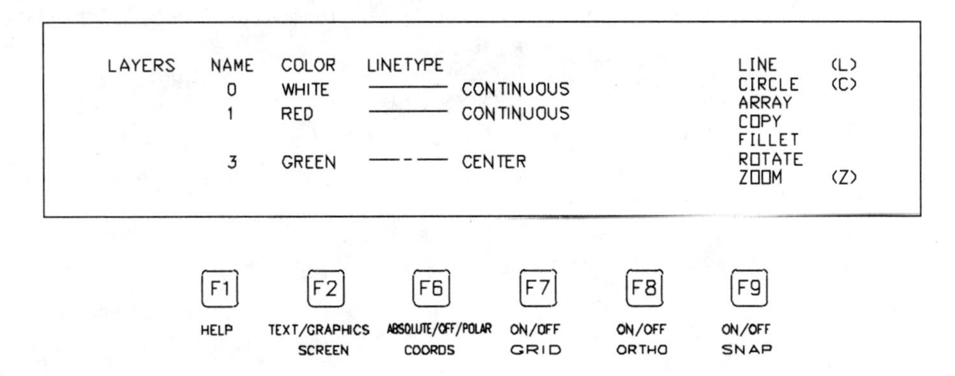

LAYERS	NAME	COLOR	LINETYPE	
	0	WHITE	————	CONTINUOUS
	1	RED	————	CONTINUOUS
	3	GREEN	—·–·—	CENTER

LINE (L)
CIRCLE (C)
ARRAY
COPY
FILLET
ROTATE
ZOOM (Z)

F1	F2	F6	F7	F8	F9
HELP	TEXT/GRAPHICS SCREEN	ABSOLUTE/OFF/POLAR COORDS	ON/OFF GRID	ON/OFF ORTHO	ON/OFF SNAP

DRAWING 5-2: ALIGNMENT WHEEL

This drawing shows a typical use of the MIRROR command. Carefully mirroring sides of the symmetrical front view will save you from duplicating some of your drawing efforts. Notice that you will need a small snap setting to draw the vertical lines at the chamfer.

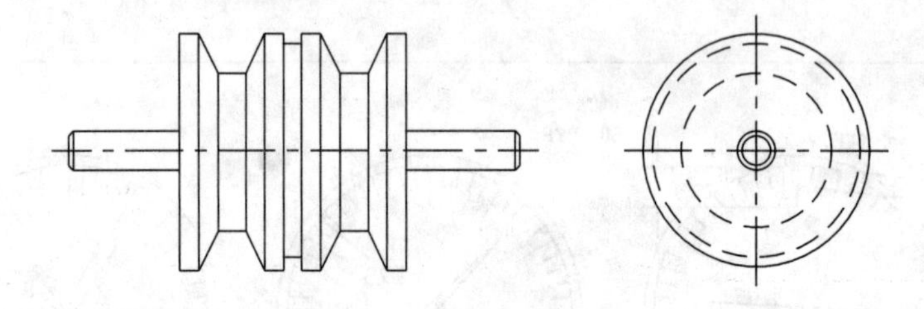

DRAWING SUGGESTIONS

GRID = .25 LTSCALE = .50
SNAP = .125 LIMITS = (0,0) (12,9)

> There are numerous ways to use MIRROR in drawing the front view. As the reference shows, there is top–bottom symmetry as well as left–right symmetry. The exercise for you is to choose an efficient mirroring sequence.

> Whatever sequence you use, consider the importance of creating the chamfer and the vertical line at the chamfer before this part of the object is mirrored.

> Once the front view is drawn, the right side view will be easy. Remember to change layers for center and hidden lines and to line up the small inner circle with the chamfer.

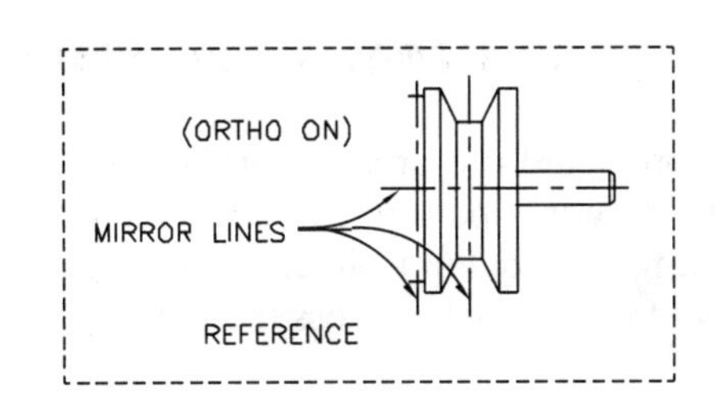

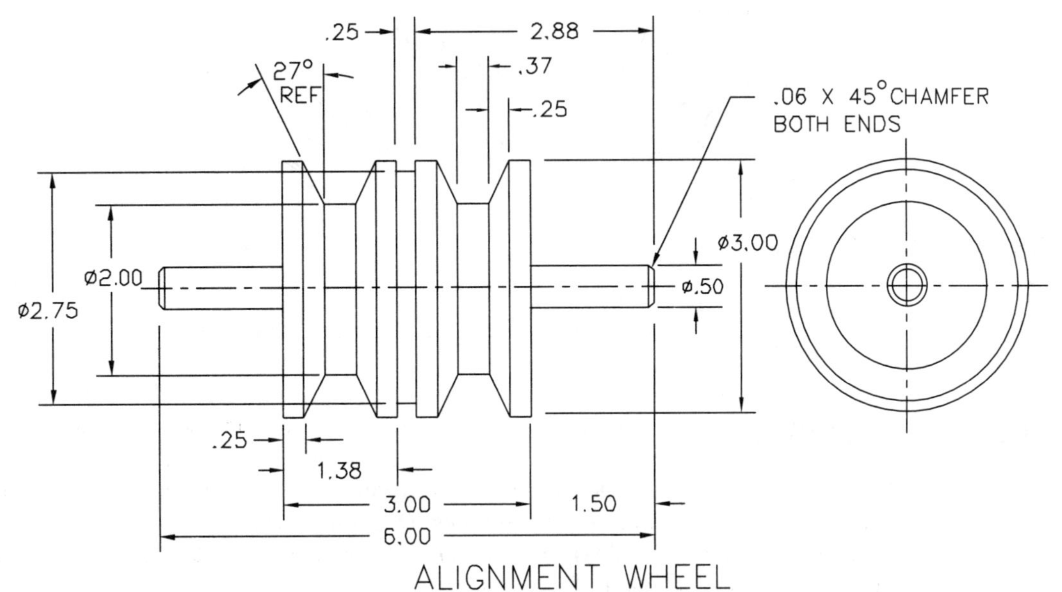

Drawing 5–2

LAYERS

NAME	COLOR	LINETYPE
0	WHITE	CONTINUOUS
1	RED	CONTINUOUS
2	YELLOW	HIDDEN
3	GREEN	CENTER

LINE (L)
CIRCLE (C)
CHAMFER
MIRROR
ZOOM (Z)

F1	F2	F6	F7	F8	F9
HELP	TEXT/GRAPHICS SCREEN	ABSOLUTE/OFF/POLAR COORDS	ON/OFF GRID	ON/OFF ORTHO	ON/OFF SNAP

DRAWING 5-3: HEARTH

Once you have completed this architectural drawing as it is shown, you might want to experiment with filling in a pattern of firebrick in the center of the hearth. The drawing itself is not complicated, but little errors will become very noticeable when you try to make the row of 4 × 8 bricks across the bottom fit with the arc of bricks across the top, so work carefully.

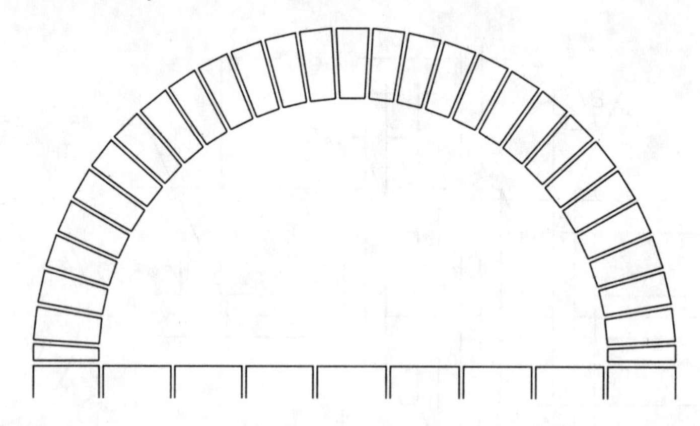

DRAWING SUGGESTIONS

UNITS = Architectural
smallest fraction = 8 (1/8″)
LIMITS = (0,0) (12′,9′)
GRID = 1′
SNAP = 1/8″

> Zoom in to draw the wedge-shaped brick indicated by the arrow on the right of the dimensioned drawing. Draw half of the brick only and mirror it across the centerline as shown. (Notice that the centerline is for reference only.) It is very important that you use MIRROR so that you can erase half of the brick later.
> Array the brick in a 29 item, 180 degree polar array.
> Erase the bottom halves of the end bricks at each end.
> Draw a new horizontal bottom line on each of the two end bricks.
> Draw a 4 × 8 brick directly below the half brick at the left end.
> Array the 4 × 8 brick in a 1 row, 9 column array, with 8.5″ between columns.

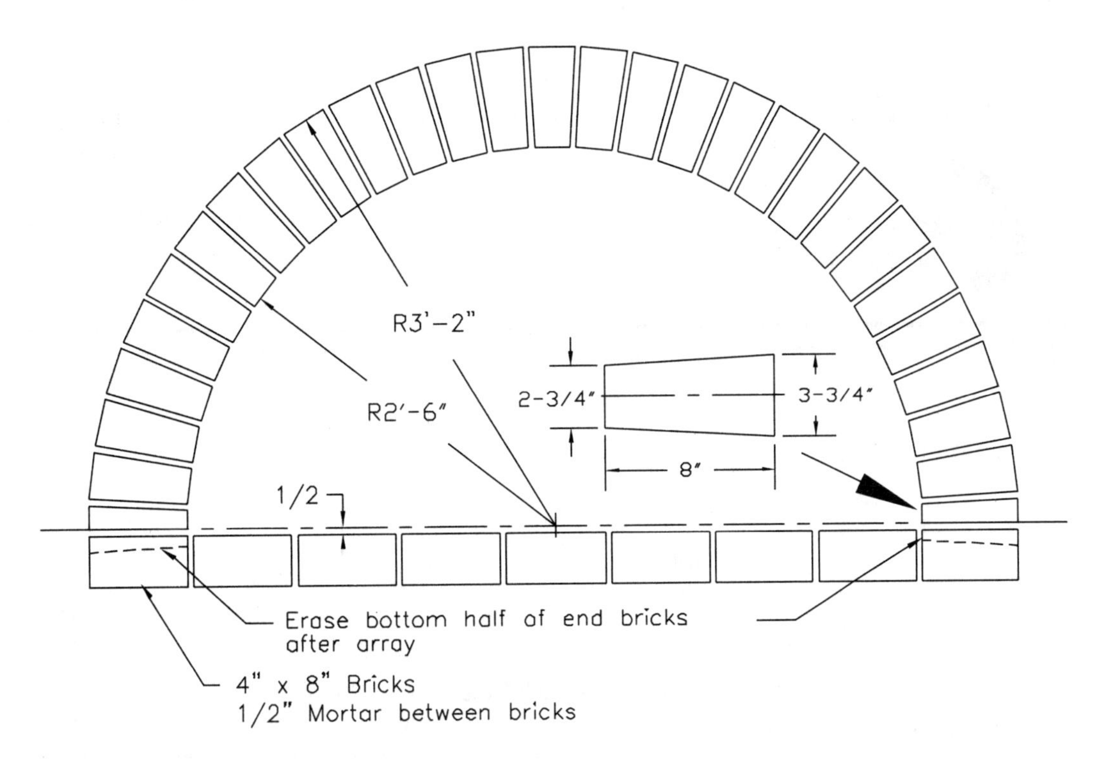

HEARTH

Drawing 5-3

Drawing Compliments of Thomas Casey

LAYERS	NAME	COLOR	LINETYPE	
	0	WHITE	———	CONTINUOUS
	1	RED	———	CONTINUOUS
	3	GREEN	— — —	CENTER

LINE (L)
ERASE (E)
MIRROR
ARRAY
MOVE (M)
ZOOM (Z)

F1	F2	F6	F7	F8	F9
HELP	TEXT/GRAPHICS SCREEN	ABSOLUTE/OFF/POLAR COORDS	ON/OFF GRID	ON/OFF ORTHO	ON/OFF SNAP

CHAPTER 6

COMMANDS

MODIFY
BREAK
TRIM
EXTEND
SPECIAL TOPIC: Object Snap (OSNAP)

OVERVIEW

This chapter will continue to expand your repertoire of editing commands. You will learn to BREAK entities on the screen into pieces so that they may be manipulated separately, or so that you can erase parts. You will also learn to shorten entities using the TRIM command, or to lengthen them with the EXTEND command.

Most important, you will begin to use a very powerful tool called Object Snap that will take you to a new level of accuracy and efficiency as a CAD operator.

TASKS

1. Use object snap to select specifiable points on an entity using single-point overrides.
2. Select points with OSNAP using running modes.
3. Use BREAK to break a previously drawn entity into two separate entities.
4. Use TRIM to shorten entities.
5. Use EXTEND to lengthen entities.
6. Plot from Multiple Viewports in Paper Space.
7. Do Drawing 6-1 ("Archimedes Spiral").
8. Do Drawing 6-2 ("Spiral Designs").
9. Do Drawing 6-3 ("Grooved Hub").

TASK 1: Selecting Points With Object Snap (Single Point-Override)

Some of the drawings in the last two chapters have pushed the limits of what you can accomplish accurately on a CAD system with incremental snap alone. Object snap is a related tool that works in a very different manner. Instead of snapping to points defined by the coordinate system, it snaps to geometrically specifiable points on objects that you already have drawn.

Let's say you want to begin a new line at the end point of one that is already on the screen. If you are lucky it may be on a snap point, but it is just as likely not to be. Turning snap off and using the arrow keys may appear to work, but chances are that when you zoom in you will find that you have actually missed the point. Using object snap is the only precise way, and it is as precise as you could want. Let's try it.

> To prepare for this exercise, draw a 6 × 6 box with a circle inside, as in *Figure 6-1*. Exact sizes and locations are not important; however, the circle should be centered within the square.
> Now enter the LINE command (type "l" or select the Line tool).

We are going to draw a line from the lower left corner of the square to a point on a line tangent to the circle, as shown in *Figure 6-2*. Notice that this task would be extremely difficult without object snap. The corner is easy to locate, since you probably have drawn it on snap, but the tangent may not be.

We will use an "end point" object snap to locate the corner and a "tangent" object snap to locate the tangent point.
> At the "From point:" prompt, instead of specifying a point, type "end" or select the Snap to Endpoint tool from the Object Snap flyout on the Standard toolbar, as illustrated in *Figure 6-3*.

You may also have a cursor menu which can be accessed by holding down the Shift key and pressing the enter (right) button on a two-button mouse.

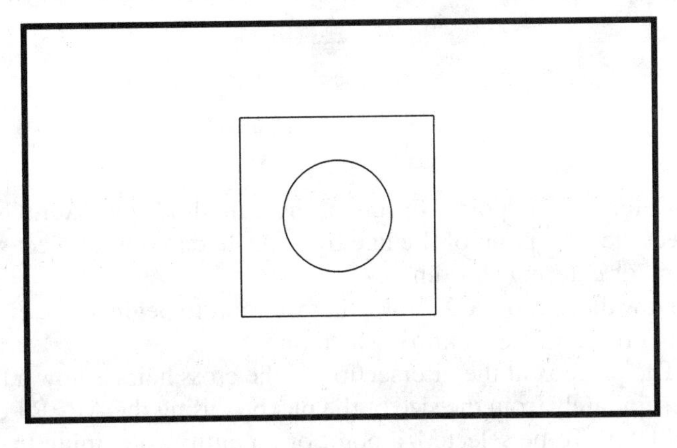

Figure 6-1

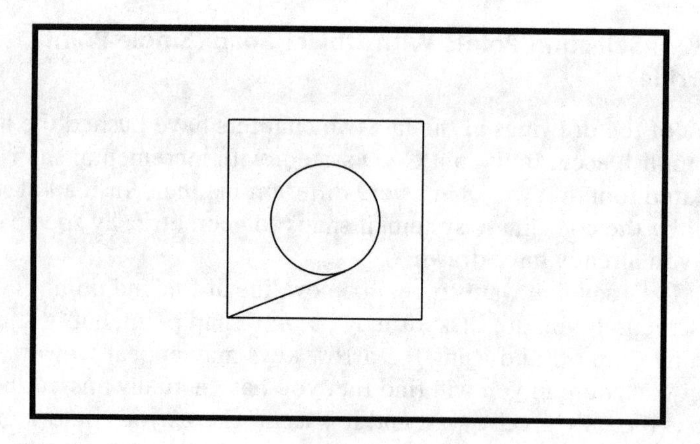

Figure 6-2

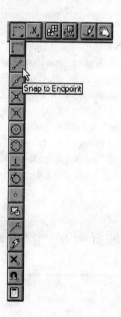

Figure 6-3

Entering "Endpoint" by any of these methods tells AutoCAD that you are going to select the start point of the line by using an end point object snap rather than direct pointing or entering coordinates.

Now that AutoCAD knows that we want to begin at the end point of a previously drawn entity, it needs to know which one.

The pickbox at the intersection of the cross hairs is now a target box. Its size can be set separately from the size of the pickbox using the APERTURE command, as discussed later. To be selected, a point or an entity containing the point must be within the aperture, as in *Figure 6-4*.

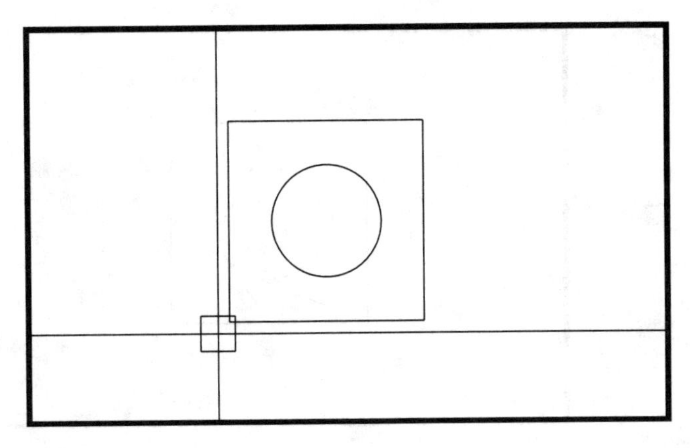

Figure 6-4

> Position the cursor so that the lower left corner of the square is within the target box, then press the pick button.

Now we will draw the tangent.

> At the "To point:" prompt, type "tan" or select "Tangent" or the Tangent tool.

> Move the cursor to the right and position the cross hairs so that the circle crosses the target box. Press the pick button. AutoCAD will locate the tangent point and draw the line.

> Press enter to exit the LINE command. Your screen should now resemble *Figure 6-2.*

We will repeat the process now, but start from the midpoint of the bottom side of the square instead of its end point.

> Repeat the LINE command.

> At the prompt for a point, type "mid" or select "Midpoint" or the Midpoint tool.

> Position the aperture anywhere along the bottom side of the square and press the pick button.

> At the prompt for a second point, type "tan" or select "Tangent" or the Tangent tool.

> Position the aperture along the lower right side of the circle and press the pick button.

> Press enter or the space bar to exit the LINE command.

At this point, your screen should resemble *Figure 6-5.*

That's all there is to it. Remember the steps: 1) enter a command; 2) when AutoCAD asks for a point, type or select an Object snap mode instead; 3) select an object to which the mode can be applied and AutoCAD will find the point.

TASK 2: Selecting Points with OSNAP (Running Mode)

So far we have been using object snap one point at a time. Since object snap is not constantly in use for most applications, this single point method is probably most common. If you find that you are going to be using one or a number of object snap types repeatedly and will not need to select many points without them, there is a way to keep

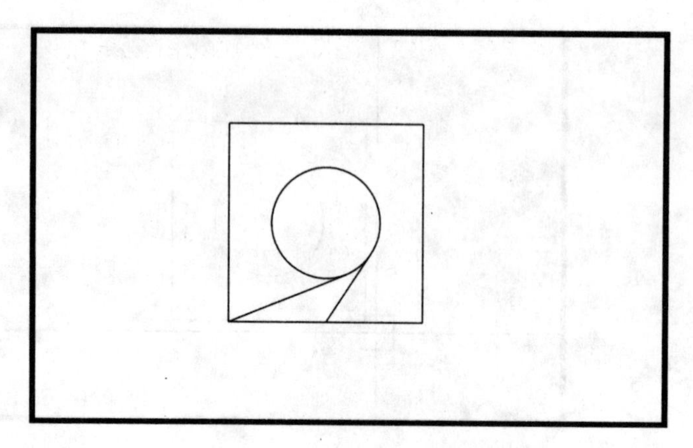

Figure 6-5

object snap modes on so that they affect all point selections. These are called "running object snap modes". We will use this method to complete the drawing shown in *Figure 6-6.* Notice how the lines are drawn from midpoints and corners to tangents to the circle. This is easily done with object snap.

In order to turn on a running osnap mode, you can enter the OSNAP command at the command line or use the DDOSNAP dialogue box from the pull down menu. We recommend the dialogue box because it is quick, easy, and also gives you the chance to change the aperture size.

> Select "Options" and then "Running Object Snap . . ." from the pull down.

You will see the dialogue box illustrated in *Figure 6-7.* At the top is a list of object snap modes with check boxes. At the bottom is a box with a scroll bar where you can change the object snap aperture size.

You will find a description of all of the object snap modes on the chart (*Figure 6-8*) at the end of this task, but for now we will be using three: midpoint, tangent,

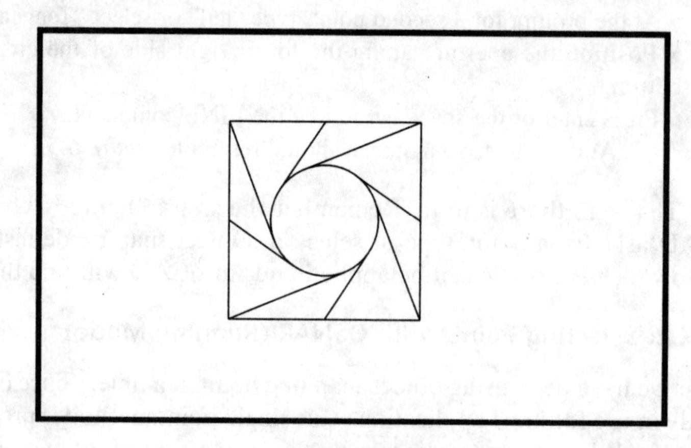

Figure 6-6

Figure 6-7

and intersect. Midpoint and tangent you already know. Intersect snaps to the point where two entities meet or cross. We will use intersect instead of end point to select the remaining three corners of the square.

> Click on the check boxes next to "Midpoint", "Tangent", and "Intersection".

Before you leave the dialogue box, try changing the aperture size.

Changing the Size of the Aperture

The aperture can be changed visually using the scroll bar in the usual manner, or by entering the APERTURE command and typing a number. If you type the command, you will be asked to specify the aperture size in pixels (from 1 to 50, one pixel being the smallest unit your monitor can display).

NOTE: The aperture size is a "system variable." This means that its setting will remain in effect when you enter other drawings.

> Click on the box in the middle of the scroll bar and drag it to the right, or click on the right arrow.

Watch the aperture in the black box on the right growing larger.

> Set the aperture by moving the scroll bar box to the left of center, as shown in *Figure 6-7*. The aperture itself should be somewhat larger than the pickbox you are used to seeing.

Remember, the aperture is distinct from the pickbox. Changing the aperture has no effect on the size of the object selection pickbox. The size of this box is controlled by another system variable called PICKBOX.

The best size for your aperture depends on your drawing. If you are doing a lot of point selection in tight spaces, and especially if you are using multiple osnap modes,

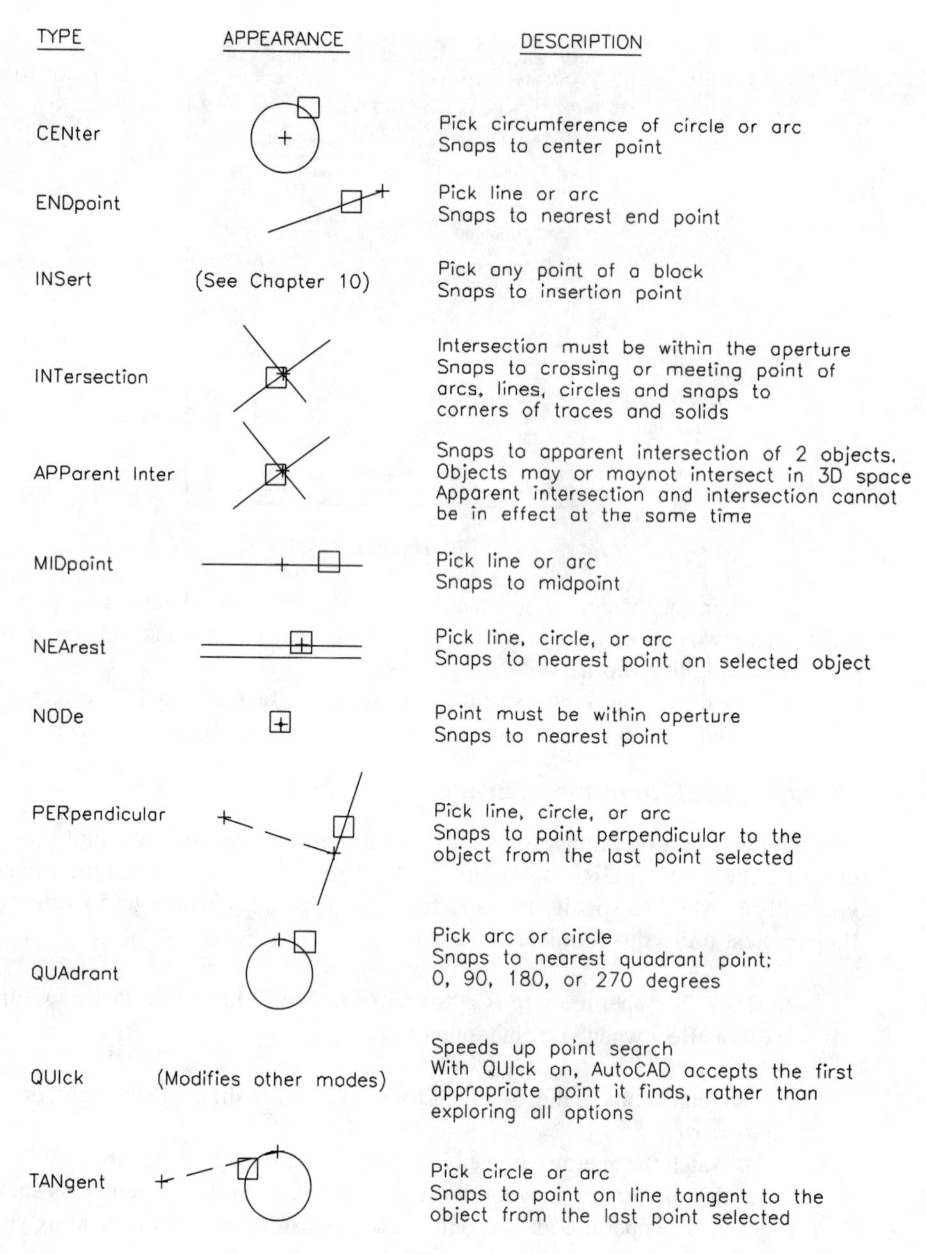

TYPE
APPEARANCE
DESCRIPTION
CENter
Pick circumference of circle or arc
Snaps to center point
ENDpoint
Pick line or arc
Snaps to nearest end point
INSert
(See Chapter 10)
Pick any point of a block
Snaps to insertion point
INTersection
Intersection must be within the aperture
Snaps to crossing or meeting point of
arcs, lines, circles and snaps to
corners of traces and solids
APParent Inter
Snaps to apparent intersection of 2 objects.
Objects may or maynot intersect in 3D space
Apparent intersection and intersection cannot
be in effect at the same time
MIDpoint
Pick line or arc
Snaps to midpoint
NEArest
Pick line, circle, or arc
Snaps to nearest point on selected object
NODe
Point must be within aperture
Snaps to nearest point
PERpendicular
Pick line, circle, or arc
Snaps to point perpendicular to the
object from the last point selected
QUAdrant
Pick arc or circle
Snaps to nearest quadrant point:
0, 90, 180, or 270 degrees
QUIck
(Modifies other modes)
Speeds up point search
With QUIck on, AutoCAD accepts the first
appropriate point it finds, rather than
exploring all options
TANgent
Pick circle or arc
Snaps to point on line tangent to the
object from the last point selected

Figure 6-8

you may want a smaller aperture to avoid confusion. If you have plenty of room to work in, you may want a larger aperture to make the selection process looser and faster.

> Click on "OK" to exit the dialogue box.

AutoCAD returns you to the "Command:" prompt and you are ready to draw with the running osnap modes in effect. We will draw tangent lines from the corners of the square and from the midpoints of their sides to produce *Figure 6-6.*

Now back to our drawing.

> Enter the LINE command.

Notice the size of the aperture on the cross hairs. The change in size tells you that you are looking at the object snap aperture instead of the pickbox and that there are osnap modes in effect. When you select points now, AutoCAD will look for the point nearest the center of the aperture that fulfills the geometric requirements of one of the three running osnap modes you have chosen. If there is more than one point, it will select the one nearest the intersection of the cross hairs.

> Position the aperture so that the lower right corner is within the box and press the pick button.

AutoCAD will select the intersection of the bottom and right sides and give you the rubber band and prompt for a second point.

Notice that the aperture is still on the cross hairs.

> Move the cross hairs up and along the right side of the circle and press the pick button.

Be sure that the intersection of the previous tangent and circle is not within the aperture. AutoCAD will construct a new tangent from the lower right corner to the circle.

> Press enter to complete the command sequence.

> Press enter again to repeat LINE so you can begin with a new start point.

We will continue to move counterclockwise around the circle. This should begin to be easy now.

> Position the aperture along the right side of the square and press the pick button. Be sure the corner is not within the aperture.

AutoCAD snaps to the midpoint of the side.

> Move up along the upper right side of the circle and press the pick button.

> Press enter to exit LINE.

> Press enter again to repeat LINE and continue around the circle-drawing tangents like this: upper right corner to top of circle, top side midpoint to top left of circle, upper left corner to left side, left side midpoint to lower left side.

Remember that running osnap modes should give you both speed and accuracy, so push yourself a little to see how quickly you can complete the figure.

Your screen should now resemble *Figure 6-6.*

Before going on, we need to turn off the running osnap modes.

> Click on "Running Object Snap . . ." on the Options pull down menu.

> Click on "Clear All" in the Running Object Snap dialogue box.

> Click on "OK" to exit the dialogue box.

AutoCAD will return you to the command prompt, and when you begin drawing again, you will see that the aperture is gone.

NOTE: You can also turn off all running osnap modes by typing "osnap" and then pressing enter when asked for osnap modes.

Now we will move on to three very useful and important new editing commands: BREAK, TRIM, and EXTEND. Before leaving Object Snap, be sure to study the chart (*Figure 6-8*).

TASK 3: BREAKing Previously Drawn Objects

The BREAK command allows you to break an object on the screen into two entities, or to cut a segment out of the middle or off the end. The command sequence is similar for all options. The action taken will depend on the points you select for breaking.

> In preparation for this section, clear your screen of any objects left over from Task 2 and draw a 5.0 horizontal line across the middle of your screen, as in *Figure 6-9*. Exact lengths and coordinates are not important. Also, be sure to turn off any running object snap modes that may be on from the last exercise.

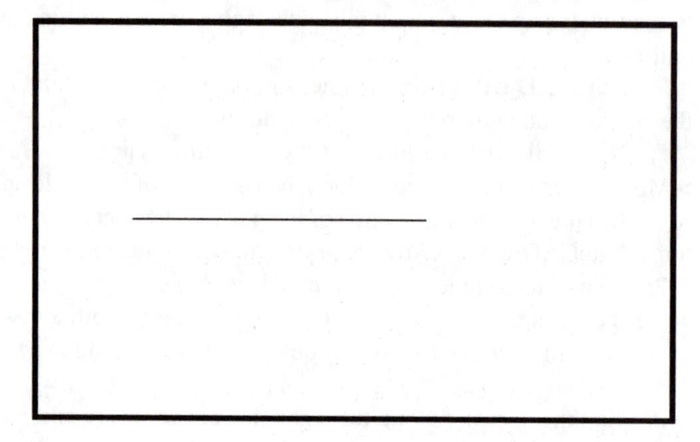

Figure 6-9

AutoCAD allows for four different ways to break an object, depending on whether the point you use to select the object is also to be considered a break point. You can break an object at one point or at two points, and you have the choice of using your object selection point as a break point or not.

We begin by breaking the line you have just drawn into two independent lines, using a single break point that is also the point used to select the line.

> Type "Break" or select the Point tool from the Modify toolbar, shown in *Figure 6-10*.
Be aware that the noun/verb or pick first sequence does not work with BREAK.

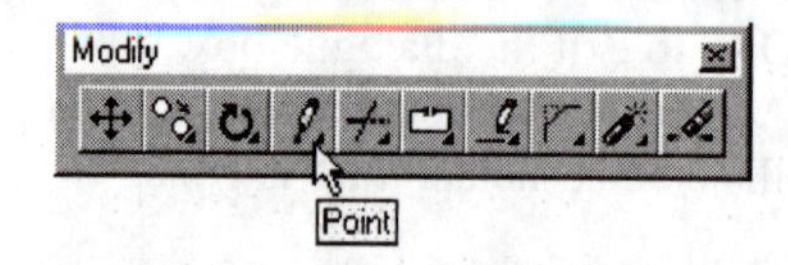

Figure 6-10

AutoCAD will prompt you to select an object to break:

Select object:

You may select an object in any of the usual ways, but notice that you can only break one object at a time. If you try to select more, with a window, for example, AutoCAD will give you only one. Because of this, you will best indicate the object you want to break by pointing to it.

NOTE: Object snap modes work well in edit commands like BREAK. If you wish to break a line at its midpoint, for example, you can use the midpoint object snap mode to select the line and the break point.

> Select the line by picking any point near its middle. (The exact point is not critical; if it were, we could use a midpoint object snap.)

The line has now been selected for breaking, and since there can be only one object, you do not have to press enter to end the selection process as you often do in other editing commands.

The break is complete. In order to demonstrate that the line is really two lines now, we will select the right half of it for our next break.

> Press enter or the space bar to repeat the BREAK command.
> Point to the line on the right side of the last break.

The right side of the line should become dotted, as in *Figure 6-11.* Clearly the line is now being treated as two separate entities.

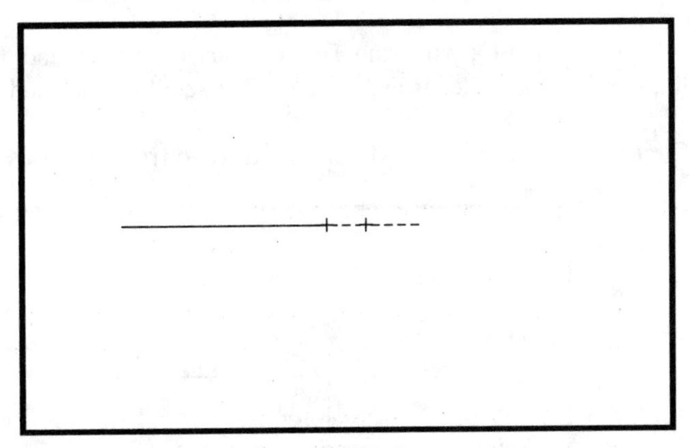

Figure 6-11

We will shorten the end of this dotted section of the line. Assume that the point you just used to select the object is the point where you want it to end; now all you need to do is to select a second point anywhere beyond the right end of the line.

> Select a second point beyond the right end of the line.

Your line should now be shortened, as in *Figure 6-12.*

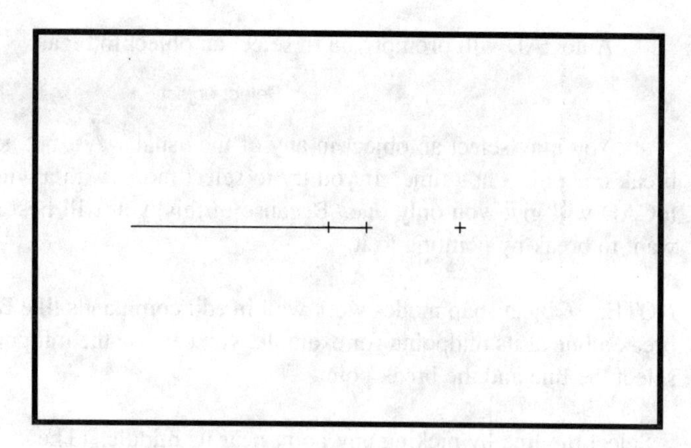

Figure 6-12

BREAK is a very useful command, but there are times when it is cumbersome to shorten objects one at a time. The TRIM command has some limitations that BREAK does not have, but it is much more efficient in situations where you want to shorten objects at intersections.

TASK 4: Using the TRIM Command

The TRIM command works wonders in many situations where you want to shorten objects at their intersections with other objects. The only limitation is that you must have at least two objects and they must cross or meet. If you are not trimming to an intersection, use BREAK.

> In preparation for exploring TRIM, clear your screen and then draw two horizontal lines crossing a circle, as in *Figure 6-13*. Exact locations and sizes are not important.

First we will use the TRIM command to go from *Figure 6-13* to *Figure 6-14*.

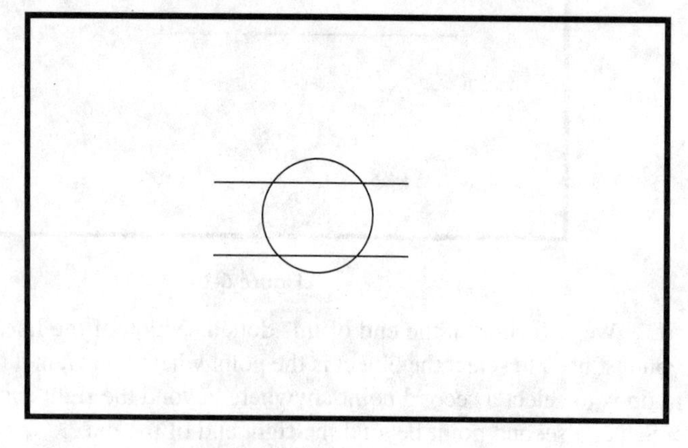

Figure 6-13

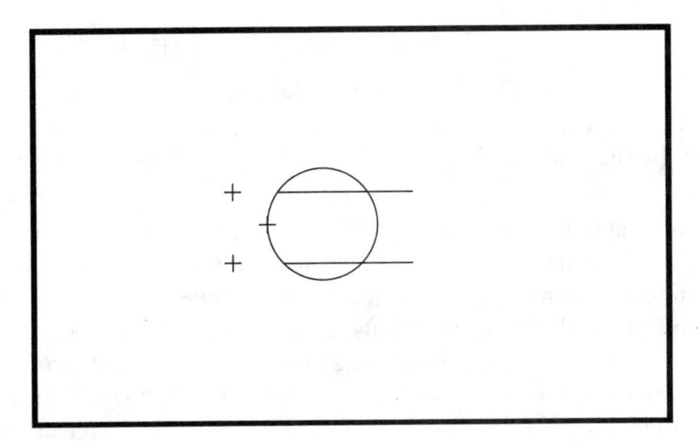

Figure 6-14

> Type "Trim" or select the Trim tool from the Trim flyout on the Modify toolbar, as illustrated in *Figure 6-15.*

The first thing AutoCAD will want you to specify is at least one cutting edge. A cutting edge is an entity you want to use to trim another entity. That is, you want the trimmed entity to end at its intersection with the cutting edge. You may have to press F2 to see the complete prompt, which looks like this:

```
Select cutting edge(s): (Projmode=UCS, Edgemode=No extend)
Select objects:
```

The first line reminds you that you are selecting edges first—the objects you want to trim will be selected later. The option of selecting more than one edge is a useful one, which we will get to shortly. The notations in parentheses are relevant to 3D drawing. Projmode and Edgemode are system variables that determine the way AutoCAD will interpret boundaries and intersections in 3D space.

For now, we will select the circle as an edge and use it to trim the upper line.

> Point to the circle.

The circle becomes dotted and will remain so until you leave the TRIM command. AutoCAD will prompt for more objects until you indicate that you are through selecting edges.

> Press enter or the space bar to end the selection of cutting edges.

You will be prompted for an object to trim:

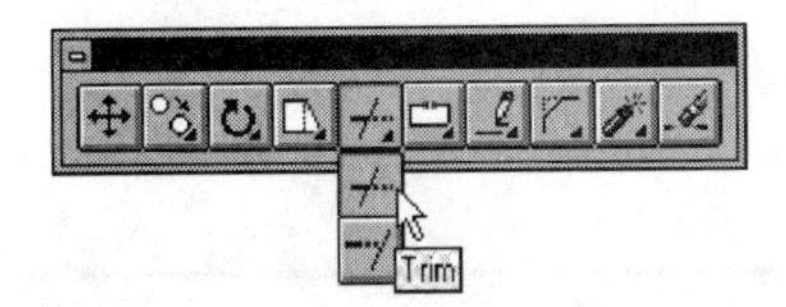

Figure 6-15

<Select object to trim>/Project/Edge/Undo:

We will trim off the segment of the upper line that lies outside the circle on the left. The important thing is to point to the part of the object you want to remove, as shown by the blips in *Figure 6-14*.

> Point to the upper line to the left of where it crosses the circle.

The line is trimmed immediately but the circle is still dotted, and AutoCAD continues to prompt for more objects to trim. Note how this differs from the BREAK command, in which you could only break one object at a time.

Also notice that you have an undo option, so that if the trim does not turn out the way you wanted, you can back up without having to leave the command and start over.

> Point to the lower line to the left of where it crosses the circle.

Now you have trimmed both lines.

> Press enter or the space bar to end the TRIM operation.

Your screen should resemble *Figure 6-14*.

This has been a very simple trimming process, but more complex trimming is just as easy. The key is that you can select as many edges as you like and that an entity may be selected as both an edge and an object to trim, as we will demonstrate.

> Repeat the TRIM command.

> Select both lines and the circle as cutting edges. This can be done with a window or with a crossing box.

> Press enter to end the selection of edges.

> Point to each of the remaining two-line segments that lie outside the circle on the right, and to the top and bottom arcs of the circle to produce the band-aid shaped object in *Figure 6-16*.

> Press enter to exit the TRIM command.

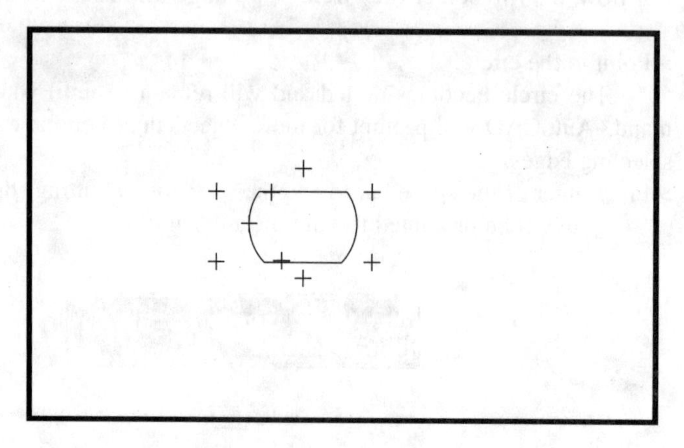

Figure 6-16

TASK 5: Using the EXTEND Command

If you compare the procedures of the EXTEND command and the TRIM command, you will notice a remarkable similarity. Just substitute the word "boundary" for "cutting edge" and the word "extend" for "trim" and you've got it. These two commands are so quick to use that it is sometimes efficient to draw a temporary cutting edge or boundary on your screen and erase it after trimming or extending.

> Leave *Figure 6-16,* the band-aid, on your screen and draw a vertical line to the right of it, as in *Figure 6-17.* We will use this line as a boundary to which to extend the two horizontal lines, as in *Figure 6-18.*

> Type "Extend" or select the Extend tool from the Trim flyout on the Modify toolbar, as shown in *Figure 6-19.*

You will be prompted for objects to serve as boundaries:

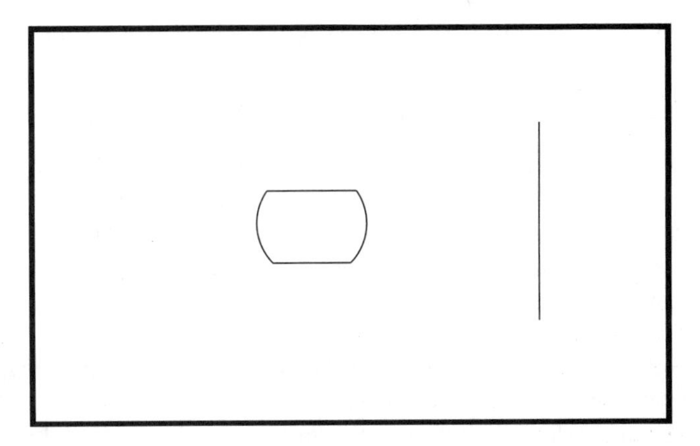

Figure 6-17

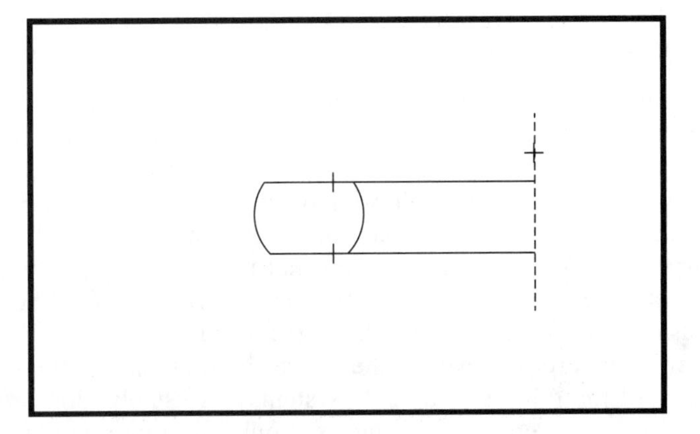

Figure 6-18

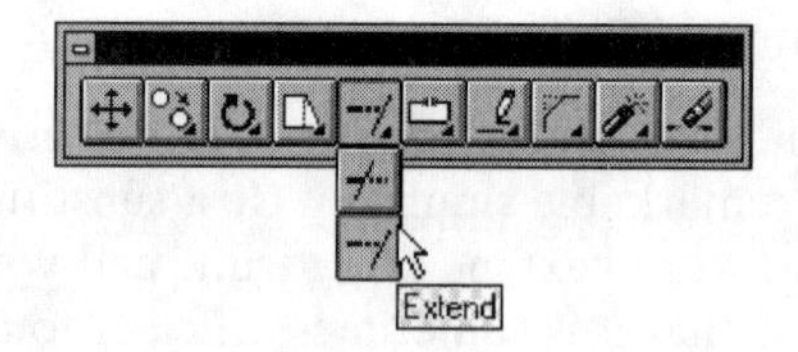

Figure 6-19

Select boundary edge(s): (Projmode=UCS, Edgemode=No extend)
Select objects:

Look familiar? As with the TRIM command, any of the usual selection methods will work. For our purposes, simply point to the vertical line.

> Point to the vertical line on the right.

You will be prompted for more boundary objects until you press enter.

> Press enter or the space bar to end the selection of boundaries.

AutoCAD now asks for objects to extend:

<Select objects to extend>/Project/Edge/Undo:

> Point to the right half of one of the two horizontal lines.

Notice that you have to point to the line on the side closer to the selected boundary. Otherwise, AutoCAD will look to the left instead of the right and give you the following message:

Object does not intersect an Edge

Note also that you can only select objects to extend by pointing. Windowing, crossing, or last selections will not work.

> Point to the right half of the other horizontal line. Both lines should be extended to the vertical line. Your screen should resemble *Figure 6-18.*

> Press enter to exit the EXTEND command.

TASK 6: Plotting from Multiple View Ports in Paper Space

In this plotting exercise, you will begin to use paper space and multiple floating viewports for the first time. We have used Drawing 6-3 to illustrate. We recommend that you read through this section now and then return to it after you complete Drawing 6-3.

You have a completed drawing on your screen and you are now considering how to present it on paper. In the days of the drafting board, all drawings were committed to paper from the start. This meant that people doing drafting were inevitably conscious of scale, paper size, and rotation from start to finish. When draftspeople first began using CAD systems, they still tended to think in terms of the final hardcopy their plotter would produce even as they were creating lines on the screen. To take full advantage of the powers of a CAD system, it is usually more efficient to conceive of screen images as representing objects at full scale, rather than as potential drawings on a piece of paper. After all, the units on a screen grid can just as easily represent miles as inches. This little world of the screen in which we can think and draw in full scale is called "model space" in AutoCAD lingo.

Using model space to its fullest potential, we can allow ourselves to ignore scale and other drawing paper issues entirely, if we wish, until it is time to plot. At that time, we may want to make use of AutoCAD's "paper space". Paper space overlays model space and allows us to create a paper-scale world on top of the full-scale world we have been drawing in.

This is the concept of paper space. In actual practice, you may find paper space unnecessary for common two-dimensional plotting. As we have seen in previous plotting exercises, the PLOT command has a very efficient previewing system and scaling features of its own that will allow you to move easily between 2D model space and the sheet of drawing paper waiting in your plotter.

Paper space really begins to pay off when you want to create multiple views of a single object and plot them all simultaneously. Without paper space and non-tiled viewports, you would have to create copies of the object and assemble them into one view. With paper space, you can create multiple views of the same object and position them wherever you like on your drawing sheet.

In this task we will create a plot of Drawing 6-3, the "Grooved Hub". The plotted drawing will include the original two views plus two close-up views. This will be done in paper space using three viewports. You should be able to generalize from the techniques demonstrated here to create a variety of multiple-view drawings with whatever drawing you may be working on. Later on, when you get into 3D drawing you will be able to draw an object and view it from different angles, placing different views in different viewports and placing these where you want them in your completed drawing.

> To begin this exercise, you should have a drawing on your screen which you want to plot showing some close-up views. We have used Drawing 6-3, shown in *Figure 6-20,* for illustration.

> Before leaving model space, turn off the grid (F7). You will see why shortly.

In order to enter paper space, you must change the value of the "Tilemode" variable from 1 ("On") to 0 ("Off"). When Tilemode is on, you can only create

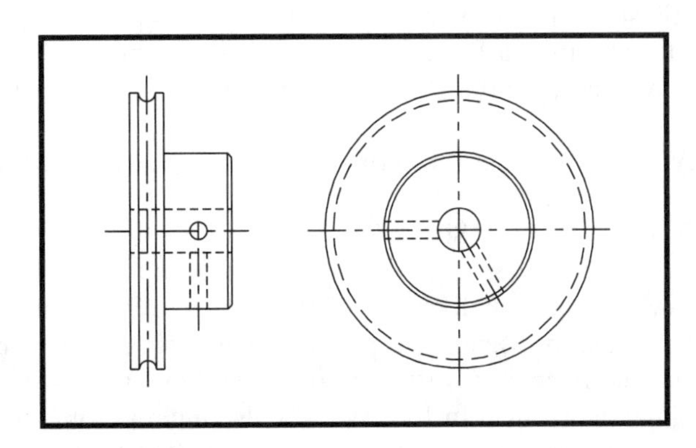

Figure 6-20

non-overlapping viewports. These viewports cannot be moved or plotted simultaneously.

You can change Tilemode at the command line, the Standard toolbar, or from the pull down menu.

> Type "Tilemode" and then "0" or select "Paper Space" from the View pull down, or the Paper Space tool from the Standard Toolbar Tilemode flyout on the Standard toolbar, as shown in *Figure 6-21*.

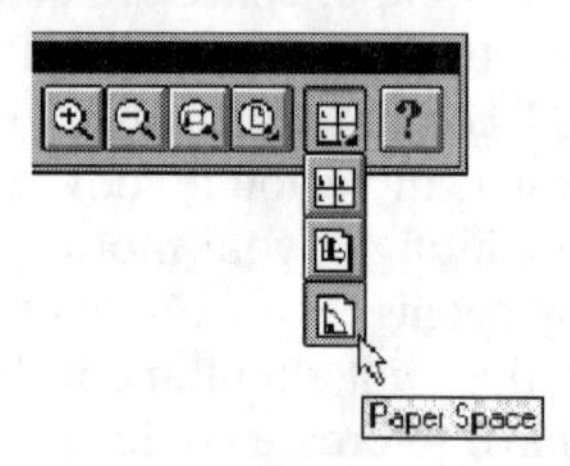

Figure 6-21

Your screen will go blank and AutoCAD will regenerate your drawing. If "Ucsicon" is "On", you will see the paper space icon in the left-hand corner of your screen. This icon appears in place of the UCS icon whenever you switch to paper space. For this reason, it is a good idea to have the icon turned on when you are going back and forth between the two spaces.

> If necessary, turn the UCS icon on by typing "ucsicon" and then "on".

You are now looking at a blank sheet of paper in paper space. You can imagine that your drawing is behind the paper space screen. We will need to create viewports before we can see it again. Viewports are like doors from paper space into model space. Until a door is opened, you cannot see what is behind it.

First, we need to think about the paper our drawing is headed for. This, after all, is the whole point of paper space. Before opening viewports, we will set up limits and a grid in paper space to emulate the drawing sheet we are going to plot on. For purposes of illustration, let's say that we are going to plot on a D-size sheet of paper. If your plotter will take a D-size sheet, you will be able to find the actual plotting area available by looking in the PLOT dialogue paper size list box.

> Type "Plot" or select the Print tool from the Standard toolbar, or "Print . . ." from the File pull down.

> Type "s" or select "Size . . .".

You will see the dialogue box illustrated previously in *Figure 5-14*. Ours shows us that the Houston Instruments plotter we are using plots in an area 33.00 by 21.00 on a D-size sheet. If yours is different, or if you have no D-size option, use a sheet size and plotting area from your list. We will also include some instructions for A-size paper as we go along since most people have access to a printer. You should expect to make adjustments, however, if your drawing size does not match ours.

> Cancel the Paper Size subdialogue.
> Cancel the Plot Configuration dialogue.
This should bring you back to your blank paper space screen.
> Type "Limits" or select "Drawing Limits . . ." from the Data pull down menu.
> Set paper space limits to (0,0) and (33,21), or whatever is appropriate for your plotter. If you are working with A size, it will be approximately (12,9) for most printers.
> Turn the grid on (F7).
The grid is now on in paper space and off in model space.
> Zoom all.
> Reset your grid and snap to a 1.00 increment.

Your screen is now truly representative of a D-size drawing sheet. The plot will be made 1 to 1, with one paper space screen unit equaling 1 inch on the drawing sheet. There is little reason to do it any other way, since the whole point of paper space is to emulate the paper on the screen.

Now it is time to create viewports.

> Type "Mview" or select "Floating Model Space" from the View pull down menu or the Floating Model Space tool from the Tilemode flyout on the Standard toolbar, as shown in *Figure 6-22*.

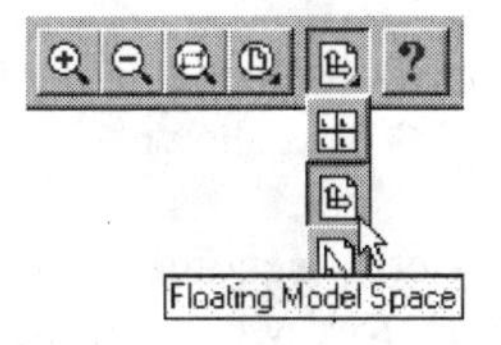

Figure 6-22

The Floating Model Space selections simply initiate the MVIEW command and return you to model space at the same time.

AutoCAD prompts:

ON/OFF/Hideplot/Fit/2/3/4/Restore/<First Point>:

We will deal only with the default option in this exercise. With this option, you create a viewport just as you would a selection window.

> Pick a point at the lower left of your screen as shown in *Figure 6-23*. Our actual point is (1.00,1.00). All of the coordinate values in this exercise will need to be adjusted if you are using something other than a D-size sheet of paper. Exact correspondence is not critical. As long as your viewports resemble our illustrations, you will be fine.
> Pick a second point up 13 and over 9 from point 1. The coordinate display will show (10.00,14.00).

Your screen will be redrawn with the drawing shown at small scale in the viewport you have just created. We will create a second viewport next.

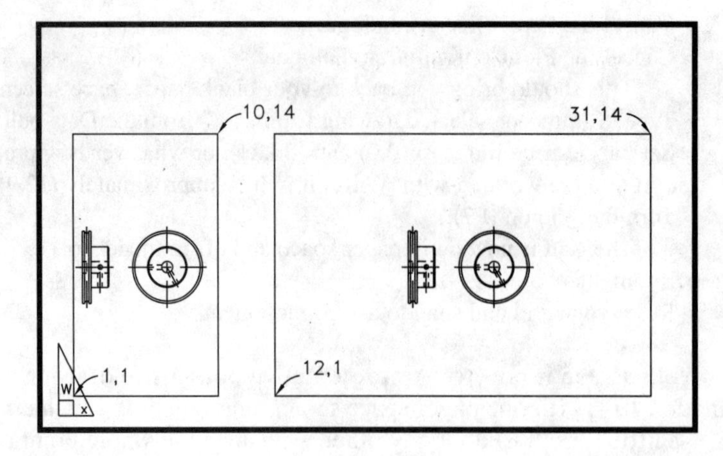

Figure 6-23

If you had not turned off the model space grid at this point, you would see model space and paper space grids overlapping. Having both grids on is useless and confusing.

Now we will create a second viewport to the right of the first.

> Repeat MVIEW.
> Pick point (12.00,1.00), as shown in the figure.
> Pick point (31.00,14.00), as shown.

You have now created a second viewport. Notice that the images in the two viewports are the same. Before we go on to create a third viewport, we will enlarge and position these to create two different images inside the viewports. To do this, we need to switch back to model space.

> If necessary, type "ms" or select "Floating Model Space" from the pull down menu or the Standard toolbar.

You will see that the paper space icon disappears and UCS icons are shown in the corners of the viewports. You will also see that the viewport on the right is outlined in black, indicating that it is currently active. Take a moment to explore the way the cursor works. You can have only one viewport active at a time. When the cursor is within that viewport, you will see the cross hairs. Otherwise you will see an arrow. To switch viewports, move the arrow into the desired viewport and press the pick button.

> Make the right viewport active.

We are going to zoom in in the right viewport so that the two-view drawing fills the viewport. We could do this by windowing, but we can maintain more precise scale relationships if we use the ZOOM "Center" and "XP" options. We will see the importance of paper space and model space scale relationships in later chapters.

> Type "z" or select the Zoom center tool from the Zoom flyout on the Standard toolbar.
> If you are typing, type "c".

AutoCAD prompts for a center point.

> Pick a point roughly halfway between the two views in the viewport.

AutoCAD will prompt for a magnification value. Accepting the default would simply center the image in the viewport. We will take an extra step by using the "XP" option here. "XP" means "times paper." It allows you to zoom relative to paper space units. If you zoom 1xp, then a model space unit will take on the size of a current paper space unit, which in turn equals one inch of drawing paper. We will zoom 2xp (.5xp if you are using A-size paper). This will mean that one unit in the viewport will equal 2 paper space units, or 2 inches on paper (in A size, 1 unit will equal .5 inches on paper).

> Type "2xp" (or .5xp for A-size paper).

Your right viewport will be redrawn to resemble the one in *Figure 6-24.* You may need to PAN slightly to the left or right to position the image in the viewport. Notice that you cannot use the scrollbars to pan inside a viewport.

Next, we will enlarge the image on the left so that we get a closeup of the left side view.

> Make the left viewport active.
> Type "z" or select the Zoom center tool.
> Type "c" or select "Center".
> Pick a center point in the middle of the left side view.
> Type "4xp" (1xp for A-size paper).

This will create an enlarged image in which one model space unit equals 4 inches in the drawing sheet (1 = 1 A size).

> Pan left or right to position the enlargement in the viewport, as shown in *Figure 6-24.*

Now that you have the technique, we will create one more enlargement, focusing on the center of the hub.

> Type "Mview" or select "Floating Viewports" from the View pull down menu and then "1 Viewport".
> Pick point (23.00,15.00).
> Pick point (31.00, 21.00).

Notice that the new viewport is created from the most recent viewport image.

> Use PAN in model space to bring the hub center into the new viewport.

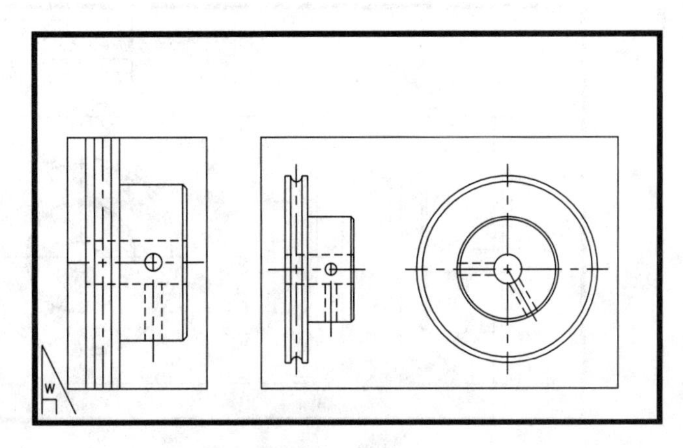

Figure 6-24

> Zoom on the center of the hub at 6 times paper (1.5xp for an A sheet) in the new viewport.
> Type or select "Regenall" to regenerate the display and create truer circles.
Your screen should resemble *Figure 6-25.*

Great! Now we are ready to plot. Plotting a multiple viewport drawing is no different from plotting from a single view, but you have to make sure you plot from paper space. Otherwise, AutoCAD will plot only the currently active viewport.

> Type "ps" or select the Paper space icon from the Tilemode flyout on the Standard toolbar.
> Type "Plot" or select the Print tool from the Standard toolbar, or "Print . . ." from the File pull down.
> Click on "Limits" in the Additional Parameters box.
> Click on "Size" in the Paper Size and Orientation box.
We looked into this box for information before, but we canceled the command. Now it is time to choose the paper size we have been preparing for.
> Select the D-size option, or whatever you have chosen for this exercise.
> Click on "OK" in the Size sub-dialogue box.
> If the Scaled to Fit box is selected, click on it so the X is removed.
When scaled to fit is de-selected, the scale boxes should show 1 = 1. If not, you should edit them to show 1 = 1.
> Do a partial preview of the plot.
The effective area should match the paper size, because we are set to plot limits and our limits match the effective drawing area of this paper size for this plotter.
> Do a full preview of the plot.
Your screen should resemble *Figure 6-26.* (We have moved the Plot Preview box to the upper left corner of the screen.)
Excellent. You are now ready to plot.

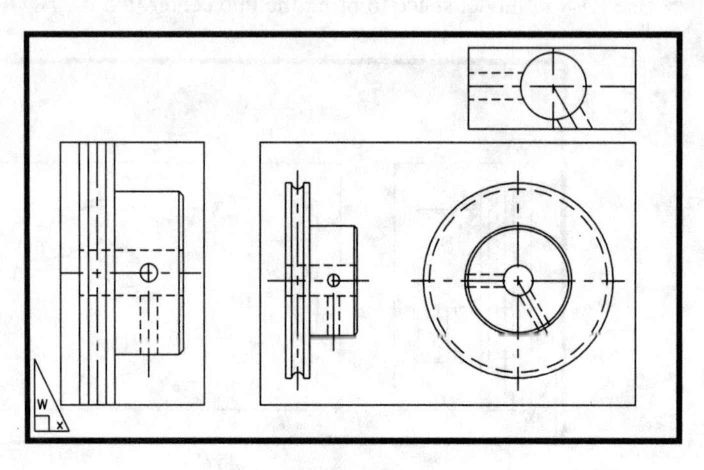

Figure 6-25

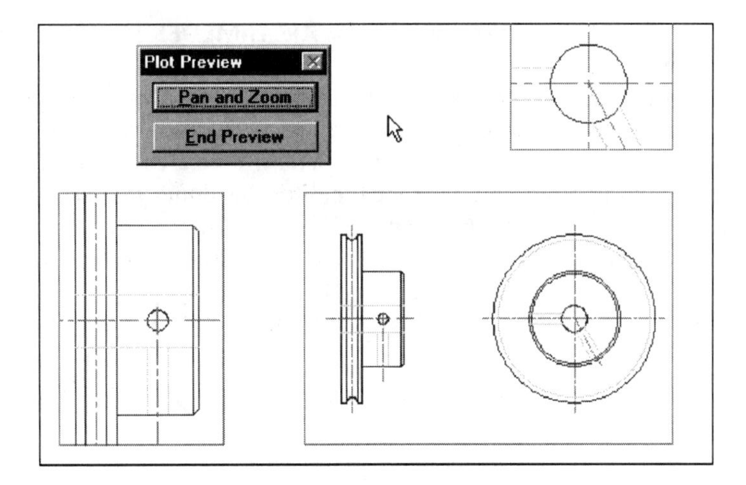

Figure 6-26

> Click on “End Preview”.
> Prepare your plotter. (Make sure you use the right size paper).
> Click on “OK”.
> Press enter and watch it go.

TASKS 7, 8, and 9

With Object snap and the new editing commands and plot procedures you have learned in this chapter, you have reached a significant plateau. Most of the fundamental drawing, editing, and printing tools are now in your repertoire of CAD skills. The three drawings that follow will give you the opportunity to practice what you have learned.

DRAWING 6-1: ARCHIMEDES SPIRAL

This drawing and the next go together as an exercise you should find interesting and enjoyable. These are not technical drawings, but they will give you valuable experience with important CAD commands. You will be creating a spiral using a radial grid of circles and lines as a guide. Once the spiral is done, you will use it to create the designs in the next drawing, 6-2, "Spiral Designs."

DRAWING SUGGESTIONS

GRID = .5   LIMITS = (0,0) (18,12)
SNAP = .25

> The alternating continuous and hidden lines work as a drawing aid. If you use different colors and layers, they will be even more helpful.
> Begin by drawing all the continuous circles on layer 0, centered near the middle of your display. Use the continuous circle radii as listed.
> Draw the continuous horizontal line across the middle of your six circles and then array it in a three-item polar array.
> Set to layer 2 for the hidden lines. The procedure for the hidden lines and circles will be the same as for the continuous lines, except the radii are different and you will array a vertical line instead of a horizontal one.
> Set to layer 1 for the spiral itself.
> Turn on a running object snap to intersection mode and construct a series of three-point arcs. Start points and end points will be on continuous line intersections; second points will always fall on hidden line intersections.
> When the spiral is complete, turn off layers 0 and 2. There should be nothing left on your screen but the spiral itself. Save it or go on to Drawing 6-2.

GROUPing Objects

Here is a good opportunity to use Release 13's GROUP command. GROUP defines a collection of objects as a single entity so that it may be selected and modified as a unit. The spiral you have just drawn will be used in the next drawing and it will be easier to manipulate if you GROUP it.

1. Type "Group" or select the Object Group tool from the Standard toolbar.
2. Type "Spiral" for a group name.
3. Click on "New <".
4. Select the six arcs with a window.
5. Press enter to end selection.
6. Click on "OK".

The spiral can now be selected, moved, rotated, and copied as a single entity.

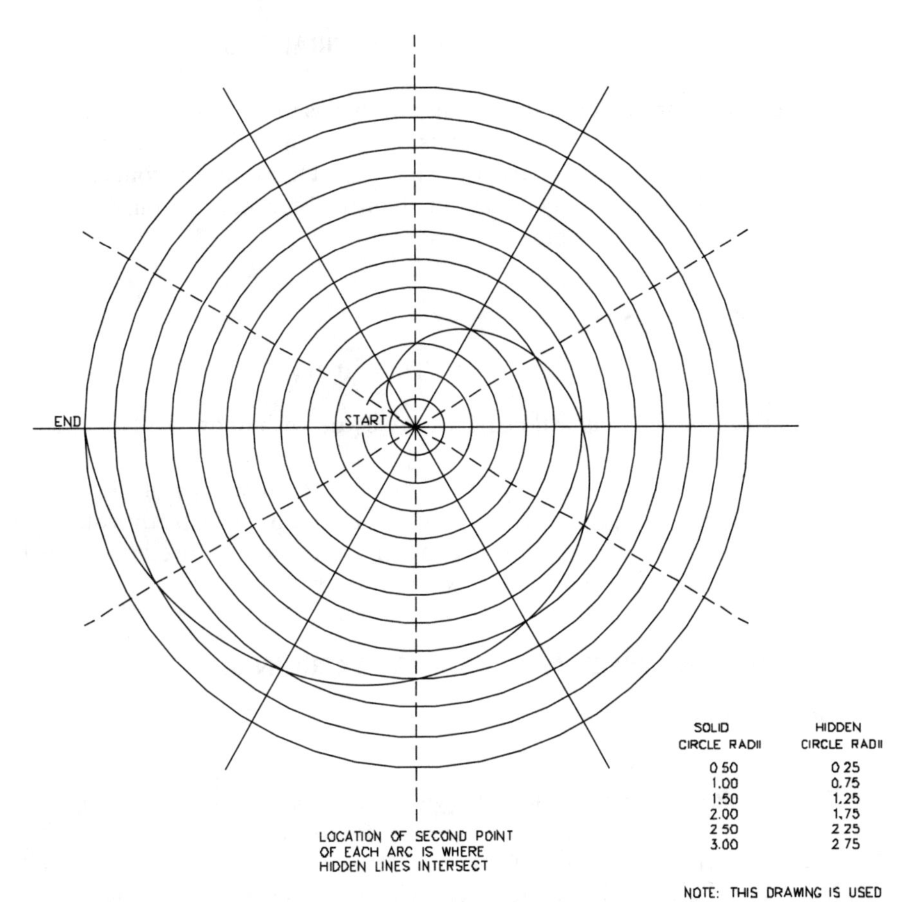

SOLID CIRCLE RADII	HIDDEN CIRCLE RADII
0.50	0.25
1.00	0.75
1.50	1.25
2.00	1.75
2.50	2.25
3.00	2.75

NOTE: THIS DRAWING IS USED ON DRAWING 6-3
SAVE THIS DRAWING!

ARCHIMEDES SPIRAL

Drawing 6–1

LAYERS	NAME	COLOR	LINETYPE	
	0	WHITE	———	CONTINUOUS
	1	RED	———	CONTINUOUS
	2	YELLOW	- - - - -	HIDDEN

LINE	(L)
CIRCLE	(C)
ARC	(A)
ARRAY	
PAN	(P)
ZOOM	(Z)

F1	F2	F6	F7	F8	F9
HELP	TEXT/GRAPHICS SCREEN	ABSOLUTE/OFF/POLAR COORDS	ON/OFF GRID	ON/OFF ORTHO	ON/OFF SNAP

DRAWING 6-2: SPIRAL DESIGNS

These designs are different from other drawings in this book. There are no dimensions and you will use only edit commands now that the spiral is drawn. Below the designs is a list of the edit commands you will need. Don't be too concerned with precision. Some of your designs may come out slightly different from ours. When this happens, try to analyze the differences.

DRAWING SUGGESTIONS

LIMITS = (0,0) (34,24)

These large limits will be necessary if you wish to draw all of these designs on the screen at once.

In some of the designs, you will need to rotate a copy of the spiral and keep the original in place. You can accomplish this by using the grip edit rotate procedure with the copy option, or by making a copy of the objects before you enter the ROTATE command. Both procedures are explained next.

HOW TO ROTATE AN OBJECT AND RETAIN THE ORIGINAL

Using Grip Edit

1. Select the spiral.
2. Pick the grip around which you want to rotate, or any of the grips if you are not going to use the grip as a base point for rotation.
3. Type "b" or select "Base point", if necessary. In this exercise, the base point you choose for rotation will depend on the design you are trying to create.
4. Type "c" or select "Copy".
5. Show the rotation angle.
6. Press enter to exit the grip edit system.

Using the ROTATE Command

1. Use COPY to make a copy of the objects you want to rotate directly on top of their originals. In other words, give the same point for the base point and the second point of displacement. When the copy is done, your screen will not look any different but there will actually be two spirals there, one on top of the other.
2. Enter ROTATE and give "p" or "previous" in response to the "Select objects:" prompt. This will select all the original objects from the last COPY sequence, without selecting the newly drawn copies.
3. Rotate as usual, choosing a base point dependent on the design you are creating.
4. After the ROTATE sequence is complete, you will need to do a REDRAW before the objects copied in the original position will be visible.

Drawing 6–2

LAYERS	NAME	COLOR	LINETYPE
	0	WHITE	———— CONTINUOUS
	1	RED	———— CONTINUOUS

LINE (L)
COPY
MOVE (M)
ROTATE
ARRAY
MIRROR
PAN (P)
ZOOM (Z)

F1	F2	F6	F7	F8	F9
HELP	TEXT/GRAPHICS SCREEN	ABSOLUTE/OFF/POLAR COORDS	ON/OFF GRID	ON/OFF ORTHO	ON/OFF SNAP

DRAWING 6-3: GROOVED HUB

This drawing includes a typical application of the rotation technique just discussed. The hidden lines in the front view must be rotated 120 degrees and a copy retained in the original position. There are also good opportunities to use MIRROR, object snap, and TRIM.

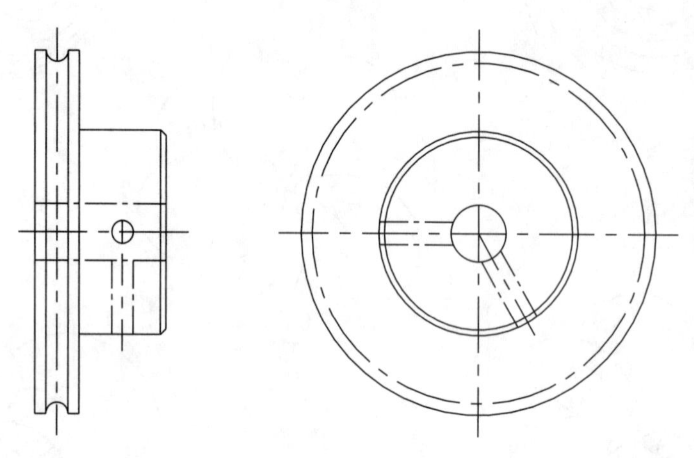

DRAWING SUGGESTIONS

GRID = .5 LIMITS = (0,0) (18,12)
SNAP = .0625

> Draw the circles in the front view and use these to line up the horizontal lines in the left side view.
> There are several different planes of symmetry in the left side view, which suggests the use of mirroring. We leave it up to you to choose an efficient sequence.
> A quick method for drawing the horizontal hidden lines in the left side view is to use a quadrant osnap to begin a line at the top and bottom of the .62 Diameter circle in the front view. Draw this line across to the back of the left side view, and use TRIM to erase the excess on both sides.
> The same method can be used to draw the two horizontal hidden lines in the front view. Snap to the top and bottom quadrants of the .25 Diameter circle in the left side view as a guide and draw lines through the front view. Then trim to the 2.25 Diameter circle and the .62 Diameter circle.
> Once these hidden lines are drawn, rotate them, retaining a copy in the original position.

Plotting in Paper Space with Multiple Viewports

This drawing is used in Task 7 to illustrate multiple viewport plotting techniques. Once you have completed the drawing as shown, save it and return to the chapter to create the drawing layout in paper space with two close-up views.

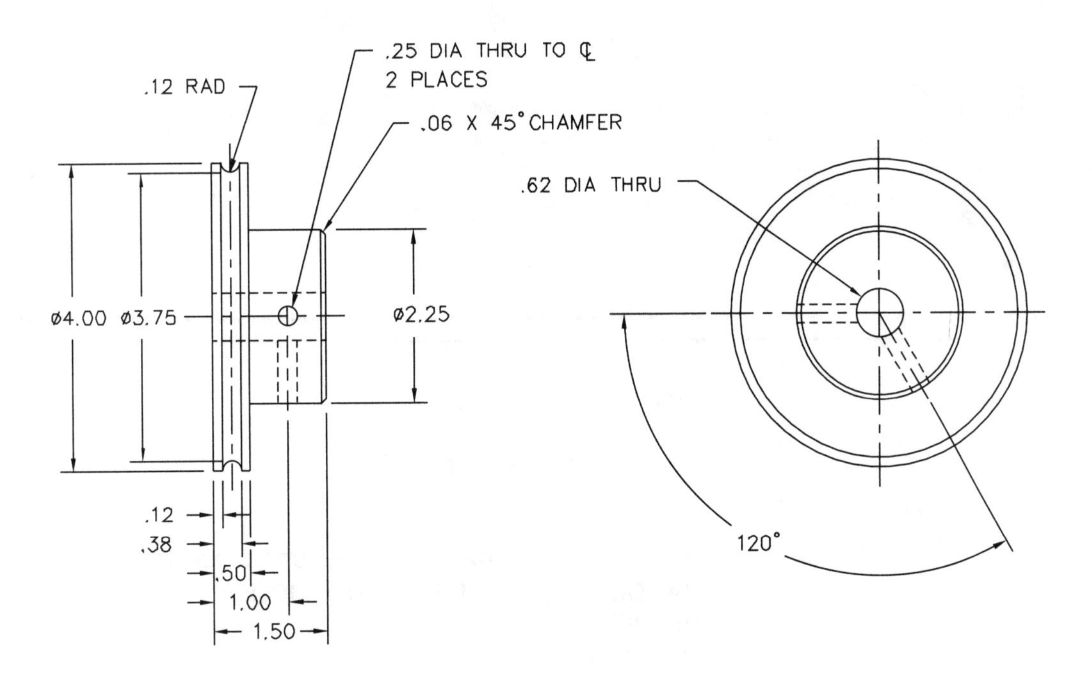

GROOVED HUB

Drawing 6-3

LAYERS	NAME	COLOR	LINETYPE	
	0	WHITE	————	CONTINUOUS
	1	RED	————	CONTINUOUS
	2	YELLOW	-------	HIDDEN
	3	GREEN	—— - —	CENTER

LINE (L)
CIRCLE (C)
ARC (A)
CHAMFER
COPY
BREAK
TRIM
MIRROR
ROTATE
ZOOM (Z)

F1	F2	F6	F7	F8	F9
HELP	TEXT/GRAPHICS SCREEN	ABSOLUTE/OFF/POLAR COORDS	ON/OFF GRID	ON/OFF ORTHO	ON/OFF SNAP

CHAPTER 7

COMMANDS

DRAW	MODIFY	DATA	OBJECT PROPERTIES
DTEXT	CHANGE	STYLE	DDEMODES
	CHPROP		
	DDCHPROP		
	DDEDIT		
	DDMODIFY		
	SCALE		

OVERVIEW

Now it's time to add text to your drawings. In this chapter you will learn to find your way around Release 13's DTEXT command. In addition, you will learn two new editing commands, CHANGE and SCALE, that are often used with text but that are equally important for editing other objects.

TASKS

1. Enter standard text using the DTEXT command.
2. Enter standard text using different justification options.
3. Use DDEDIT to edit previously drawn text.
4. Check spelling of text in a drawing.
5. Change fonts and styles.
6. Use CHANGE to edit previously drawn text.
7. Use CHANGE to edit other entities.

8. Use SCALE to change the size of objects on the screen.
9. Do Drawing 7-1 ("Title Block").
10. Do Drawing 7-2 ("Gauges").
11. Do Drawing 7-3 ("Control Panel").

TASK 1: Entering Left-Justified Text Using DTEXT

Release 13 provides three different commands for entering text in a drawing. TEXT, the oldest of the three, allows you to enter single lines of text at the command prompt. DTEXT uses all of the same options as TEXT but shows you the text on the screen as you enter it. It also allows you to enter multiple lines of text and to backspace through them to make corrections. MTEXT is a new command which allows you to type paragraphs of text in a dialogue box and then positions them in a windowed area in your drawing. All of these commands have numerous options for placing text and a variety of fonts to use and styles that can be created from them.

We will focus on the DTEXT command. Most of what you learn would transfer to the TEXT command as well, but DTEXT has more features and is easier to use.

> To prepare for this exercise, draw a 4.00 horizontal line beginning at (1,1). Then create a 6 row by 1 column array with 2.00 between rows, as shown in *Figure 7-1*. These lines are for orientation in this exercise only, they are not essential for drawing text.

> Type "Dtext" or select the Dtext tool from the Text flyout on the Draw toolbar, as shown in *Figure 7-2*.

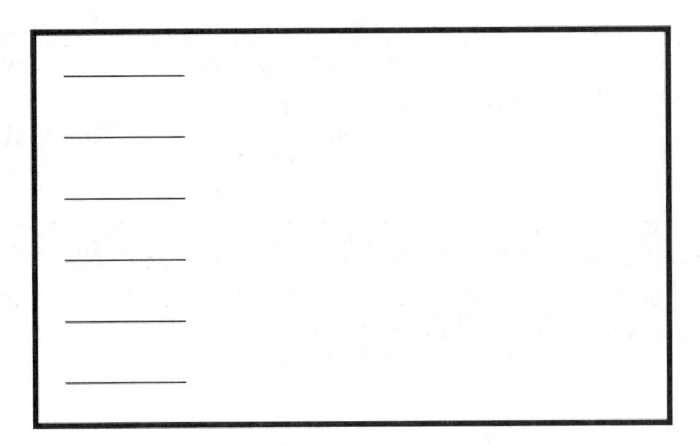

Figure 7-1

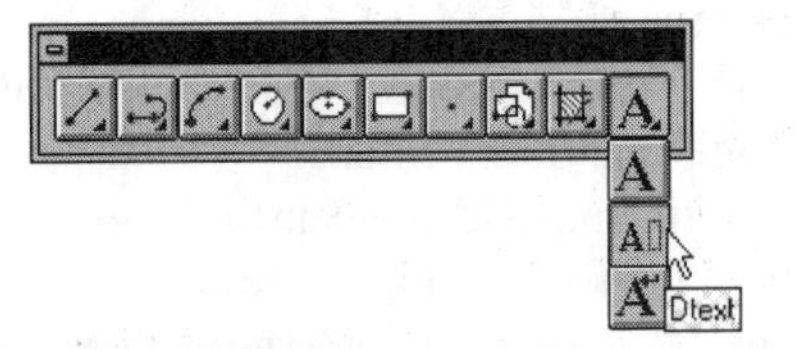

Figure 7-2

You will see a prompt with three options in the command area:

Justify/Style/<Start point>:

"Style" will be explored later. In this task, we will be looking at different options for placing text in a drawing. These are all considered text justification methods and will be listed if you choose the "Justify" option at the command prompt. We will explore these momentarily.

First, we will use the default method by picking a start point. This will give us left-justified text. This is what you would expect. The text we enter will be inserted left to right from the point we pick.

> Pick a start point at the left end of the upper line. Look at the prompt that follows and be sure that you do not attempt to enter text yet:

Height <0.20>:

This gives you the opportunity to set the text height. The number you type specifies the height of uppercase letters in the units you have specified for the current drawing. For now, we will accept the default height.

> Press enter to accept the default height (0.20).

The prompt that follows allows you to place text in a rotated position.

Rotation angle <0>:

The default of 0 degrees orients text in the usual horizontal manner. Other angles can be specified by typing a degree number relative to the polar coordinate system, or by showing a point. If you show a point, it will be taken as the second point of a baseline along which the text string will be placed. For now, we will stick to horizontal text.

> Press enter to accept the default angle (0).

Now, at last, it is time to enter the text itself. AutoCAD prompts:

Text:

Notice also that a small box has appeared at the start point. Move your cross hairs away from this point and you will see that the box stays. This shows where the first letter you type will be placed. It is a feature of DTEXT that you would not find in the TEXT command. For our text, we will type the word "Left" since this is an example of left-justified text. Watch the screen as you type and you will see dynamic text at work.

> Type "Left" and press enter. (Remember, you cannot use the space bar in place of the enter key when entering text.)

Notice that the text box jumps down below the line when you hit enter. Also notice that you are given a second "Text:" prompt in the command area.

> Type "Justified" and press enter.

The text box jumps down again and another "Text:" prompt appears. This is how DTEXT easily allows for multiple lines of text to be entered directly on the screen in a drawing. To exit the command, you need to press enter at the prompt.

> Press enter to exit DTEXT.

This completes the process and returns you to the command prompt.

Figure 7-3 shows the left-justified text you have just drawn along with the other options, as we will demonstrate in the next task.

TASK 2: Using Other Text Justification Options

We will now proceed to try out two of the other text placement options, beginning with right-justified text, as shown on the second line of *Figure 7-3*. We will also specify a change in height. The remaining three options are shown so that you can try them on your own.

Right-Justified Text

Right-justified text is constructed from an end point backing up, right to left.

> Repeat the DTEXT command. You will see the same prompt as before.
> Type "r".
Now AutoCAD prompts you for an end point instead of a start point:

End point:

We will choose the right end of the second line.
> Point to the right end of the second line.
This time, we will change the height to .50. Notice that AutoCAD gives you a rubber band from the end point. It can be used to specify height and rotation angle by pointing, if you like.
> Type ".5" and press enter or show a height of .50 by pointing.
> Press enter to retain 0 degrees of rotation. You are now prompted to enter text.
> Type "Right" and press enter.
Here you should notice an important fact about the DTEXT command. Initially, DTEXT ignores your justification choice as it enters text on the screen. The letters you type will move left to right as usual. The justification will be carried out only when you exit the command.

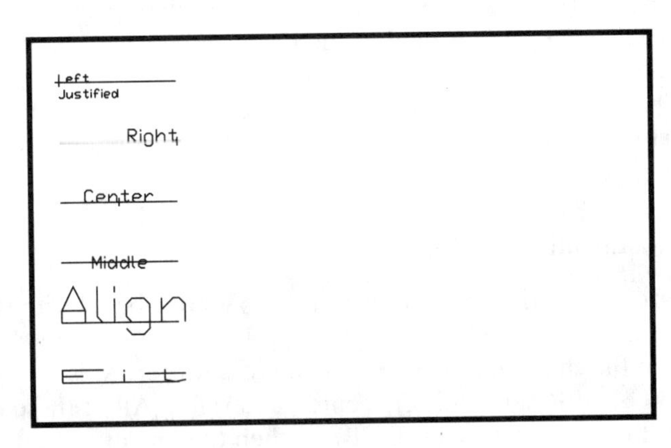

Figure 7-3

At this point, you should have the word "Right" showing to the right of the second line. Watch what happens when you press enter a second time to end the command.
> Press enter.

The text will jump to the left, back onto the line. Your screen should now include the second line of text in right-justified position as shown in *Figure 7-3*.

Centered Text

Centered text is justified from the bottom center of the text.

> Repeat the DTEXT command.
> Type "c" or select "Center".

AutoCAD prompts:

```
Center point:
```

> Point to the midpoint of the third line.
> Press enter to retain the current height, which is now set to .50.
> Press enter to retain 0 degrees of rotation.
> Type "Center" and press enter.

Notice again how the letters are typed to the screen in the usual left to right manner.
> Press enter again to complete the command.

The word "Center" should now be centered as shown in the figure.

Middle, Aligned, and Fit Text

Looking at the other three lines you will see middle text, aligned text, and text stretched to fit on the line. You can try these options using the procedures just decribed. Repeat Dtext, type in an option, and pick a point or points on the line.

Middle text (type "m") is justified from the middle of the text horizontally and vertically, instead of from the bottom.

Aligned (type "a") text is placed between two specified points. The height of the text is calculated proportional to the distance between the two points and the text is drawn along the line between the two points.

The Fit option (type "f") is similar to the Aligned option, except that the specified text height is retained. Text will be stretched horizontally to fill the line without a change in height.

Other Justification Options

There are additional justification options that are labeled with various combinations of top, middle, bottom, left, and right. The letter options are shown in *Figure 7-4*. As shown on the chart and the screen menu, T is for top, M is for middle, and B is for bottom. L, C, and R stand for left, center, and right. All of these options work the same way. Just enter TEXT or DTEXT and then the one or two letters of the option.

TEXT JUSTIFICATION	
<START POINT> TYPE ABBREVIATION	TEXT POSITION + INDICATES START POINT or PICK POINT
A	ALIGN
F	FIT
C	CENTER
M	MIDDLE
R	RIGHT
TL	TOP LEFT
TC	TOP CENTER
TR	TOP RIGHT
ML	MIDDLE LEFT
MC	MIDDLE CENTER
MR	MIDDLE RIGHT
BL	BOTTOM LEFT
BC	BOTTOM CENTER
BR	BOTTOM RIGHT

Figure 7-4

TASK 3: Editing Text with DDEDIT

There are several ways to modify text that is already in your drawing. You can change wording and spelling as well as properties such as layer, style, and justification. Commands that may be used to alter text include CHANGE, CHPROP, DDCHPROP, DDEDIT, and DDMODIFY. CHANGE, DDMODIFY, and DDEDIT can be used to change words as well as text properties; CHPROP and DDCHPROP can change only properties.

For text editing, you will most often use DDEDIT and DDMODIFY. We will begin with some simple DDEDIT text editing. In Task 5, we will use the property modification powers of DDMODIFY to show different attachment options of MTEXT paragraphs.

> Type "ddedit" or select the Edit Text tool from the Special Edit flyout on the Modify toolbar, shown in *Figure 7-5*.

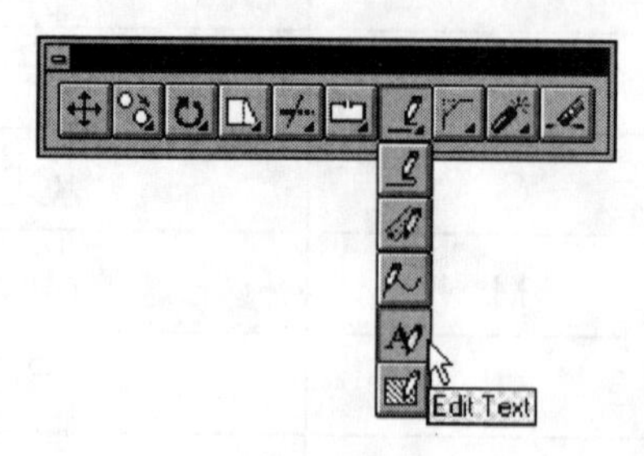

Figure 7-5

Regardless of how you enter the command, AutoCAD will prompt you to select an object for editing:

<Select an annotation object>/Undo:

Annotation objects will include all objects that have text, including text and dimensions.

We will select one line of the angled text.

> Select the words "Justified" by clicking on any of the letters.

As soon as you select the text, it will appear in a small edit box, as illustrated in *Figure 7-6*. This is the DDEDIT dialogue box for text created with DTEXT and TEXT.

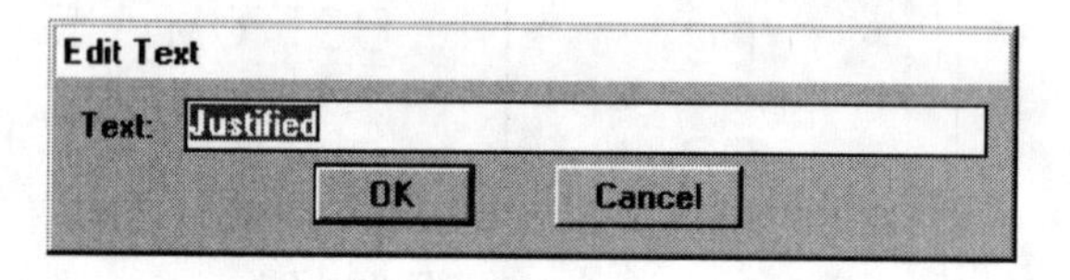

Figure 7-6

Now we will add "Dtext" to the text.

> Move the dialogue box arrow to the edit box to the right of the word "Justified" and press the pick button.

The text should no longer be highlighted and a flashing cursor should be present, indicating where text will be added if you begin typing.

> Type a space and then "Dtext" so that the line reads "Justified Dtext".

> Click on "OK".

The dialogue box will disappear and your text will be redrawn with "Dtext" added as shown in *Figure 7-7*.

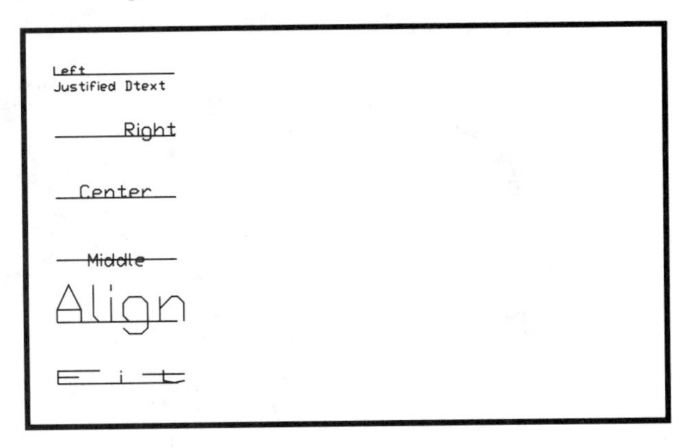

Figure 7-7

TASK 4: Using the SPELL Command

Release 13 is the first AutoCAD version to have an internal spell checker to check spelling of text in a drawing. It is simple to use and will be very familiar to anyone who has used spell checkers in word processing programs. We will use the SPELL command to check the spelling of all the text we have drawn so far.

> Type "spell" or select the Spelling tool from the Standard toolbar, as illustrated in *Figure 7-8*.

Figure 7-8

You will see a "Select objects:" prompt on the command line. At this point you could point to individual objects. Any object may be selected, although no checking will be done if you select a line, for example.

For our purposes, we will use an "All" selection to check all the spelling in the drawing.

> Type "all".

AutoCAD will continue to prompt for object selection until you press enter.

> Press enter to end object selection.

This will bring you to the Check Spelling dialogue box shown in *Figure 7-9*. If you have followed the exercise so far and not misspelled any words along the way, you will see "Dtext" in the Current word box and "Text" as a suggested correction. We will ignore this change, but before you leave SPELL, look at what is available: You can ignore a word the checker does not recognize, or change it. You can change a single instance of a word, or change all instances in the currently selected text. You can add a

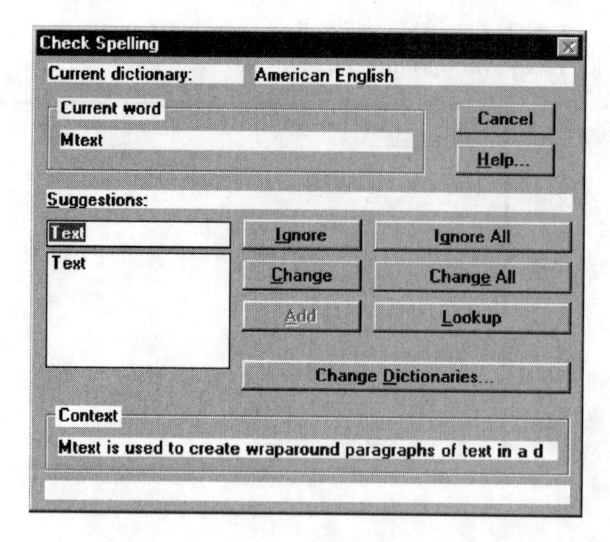

Figure 7-9

word to a customized dictionary but you cannot add words to the main dictionary. For this reason, Add is probably grayed out in your dialogue box. Selecting "Change Dictionaries . . ." will call up another dialogue box where you can create a customized dictionary and add words to it.

> Click on "Ignore".

If your drawing does not contain other spelling irregularities, you should now see an AutoCAD Alert message that says:

Spelling check complete.

> Click on "OK" to end the spell check.

If you have made any corrections in spelling, they will be incorporated into your drawing at this point.

TASK 5: Changing Fonts and Styles

By default, the current text style in any AutoCAD drawing is one called "STANDARD." It is a specific form of a font called "txt" that comes with the software. All the text you have entered so far has been drawn with the standard style of the "txt" font.

Changing fonts is a simple matter. However, there is room for confusion in the use of the words "style" and "font." You can avoid this confusion if you remember that fonts are the basic patterns of character and symbol shapes that can be used with the TEXT, DTEXT and MTEXT commands, while styles are variations in the size, orientation, and spacing of the characters in those fonts. It is possible to create your own fonts, but for most of us this is an esoteric activity. In contrast, creating your own styles is easy and practical.

We will begin by creating a vertical text style using the ROMANC (roman complex) font.

> Type "Style" or select "Text Style" from the Data pull down menu.

You will be prompted as follows:

Text style name (or ?) <STANDARD>:

By now you should be familiar with the elements of this prompt. We will use the "?" first to see a list of available styles.

> Type "?".

AutoCAD will offer you the opportunity to limit the list of styles by using wild-card characters.

Text style(s) to list <*>:

> Press enter to list all available styles.

AutoCAD will switch over to the text screen and give you the following information:

Text styles:
Style name: STANDARD Font files: txt
Height: 0.00 Width factor: 1.00 Obliquing angle: 0 Generation: Normal
Current text style: STANDARD

If anyone has used text commands in your prototype drawing, it is possible that there will be other styles listed. However, STANDARD is the only one that is certain to be there because it is created automatically. We will create our own variation of the STANDARD style and call it "VERTICAL". It will use the same "txt" character font, but will be drawn down the display instead of across.

> Press enter to repeat the STYLE command.

You will see this prompt again:

Text style name (or ?) <STANDARD>:

> Type "vertical".

AutoCAD will respond by displaying the Select Font File dialogue box shown in *Figure 7-10.*

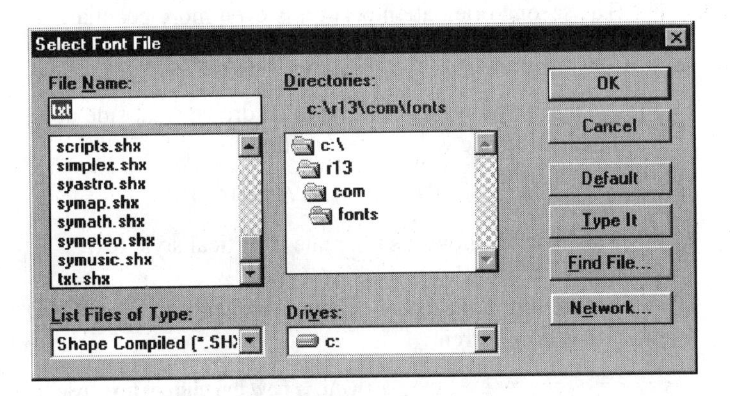

Figure 7-10

> Type "romanc" or select "romanc.shx" from the file list.
> Click on "OK".

AutoCAD closes the font dialogue box and prompts:

Height <0.00>:

It is important to understand what 0 height means in the STYLE command. It does not mean that your characters will be drawn 0.00 units high. It means that there will be no fixed height, so you will be able to specify a height whenever you use this style. Notice that VERTICAL currently has no fixed height. In general, it is best to leave the height variable as is unless you know that you will be drawing a large amount of text with one height. For practice, try giving our new "VERTICAL" style a fixed height.
> Type ".5".

AutoCAD prompts:

Width factor <1.00>:

This prompt allows you to stretch or shrink characters in the font based on a factor of 1. Let's double the width to see how it looks.
> Type "2".

AutoCAD prompts:

Obliquing angle:

This allows you to put any font on a slant, right or left, creating an italic effect. We will leave this one alone for the moment.
> Press enter to retain 0 degrees of slant.

AutoCAD follows with a series of three prompts regarding the orientation of characters. The first one is:

Backwards? <N>:

Obviously this is for special effects. You can try this one later, if you like.
> Press enter to retain "frontwards" text.

The second orientation prompt is even more peculiar:

Upside down? <N>:

> Press enter if you want your text to be drawn right side up.

Finally, the one we've been waiting for:

Vertical? <N>:

This is what allows us to create a vertical style text.
> Type "y".

Before returning to the "Command:" prompt, AutoCAD will tell you that your new style is now current:

VERTICAL is now the current text style.

To see your new style in action, you will need to enter some text.

> Type "Dtext" or select the Dtext tool from the Draw toolbar.

> Pick a start point, as shown by the blip near the letter "V" in *Figure 7-11.*

Notice that you are not prompted for a height because the current style has height fixed at .50.

> Press enter to retain 270 degrees of rotation.

> Type "Vertical".

> Press enter to end the line.

Before going on, notice that the DTEXT text placement box has moved up to begin a new column of text next to the word vertical.

> Press enter to exit DTEXT.

Your screen should resemble *Figure 7-11.*

Switching the Current Style

There are now two styles defined in your drawing. All new text is created in the current style. The style of previously drawn text can be changed, as we will show later. In this exercise, we will talk only about switching the current style.

Once you have a number of styles defined in a drawing, you can switch from one to another with the Style option of the TEXT and DTEXT commands. This option is only for switching previously defined styles; it will not allow you to define new ones. In addition, you can switch styles using the Select Text Style dialogue box. This is a subdialogue of the DDEMODES command. It can be reached by typing "DDemodes", selecting the DDEMODES tool from the Object Properties toolbar, or selecting "Object Creation . . ." from the Data pull down menu. We will use the menu.

> Select "Object Creation . . ." under "Data" on the pull down menu.

This brings up the Object Creation Modes dialogue box shown in *Figure 7-12.*

> Click on "Text Style . . .".

This will open the Select Text Style box shown in *Figure 7-13.* Vertical will be highlighted as the current text style, and will be listed along with standard and vertical.

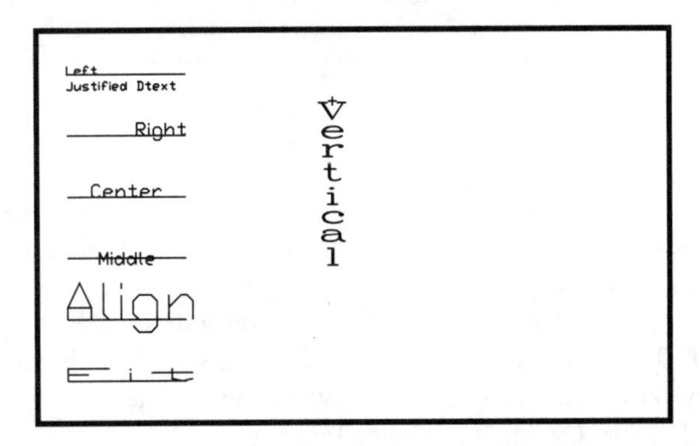

Figure 7-11

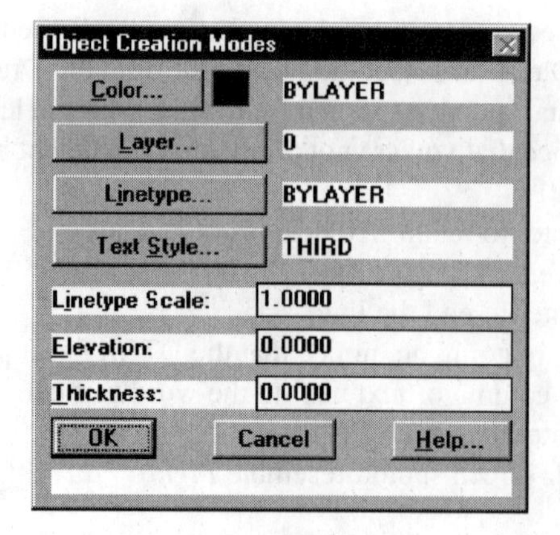

Figure 7-12

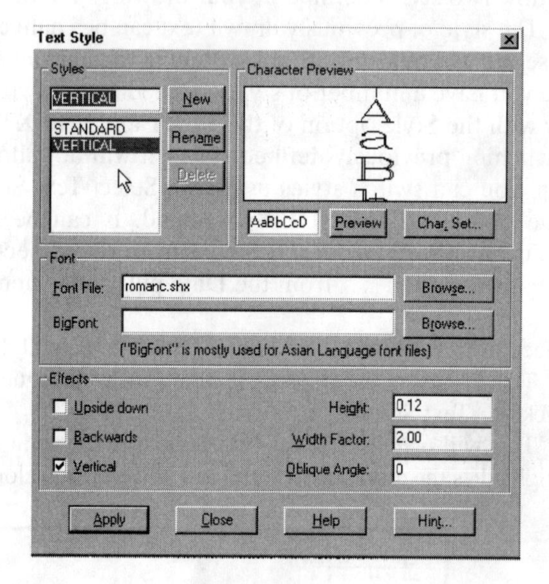

Figure 7-13

A sample of the selected text style will be shown in the black box to the right. The sample text will be ABC, but you can change this by typing in the Sample Text edit box.
> Click on "Standard".

When Standard is highlighted, you will see the sample box change to standard text. This is a convenient way to view and switch among text styles. Remember, it can only show you styles that you have already created.
> Click on "OK" to exit Select Text Style.
> Click on "OK" to exit Object Creation Modes.

NOTE: If you change the definition of a text style, all text previously drawn in that style will be regenerated with the new style specifications.

TASK 6: Changing Previously Drawn Text With CHANGE

The CHANGE command is another useful tool that allows you to change a number of different entities, text being one of them. CHANGE works from the command line and will allow most of the same changes that you will also find in DDMODIFY. The main advantage of CHANGE is that you can select more than one object and work through the changes without leaving the command. The qualities of previously drawn text that can be changed are location, style, height, rotation angle, and the text itself. You can change all the properties at once if need be, but more often you will be changing only one or two. Properties you do not wish to change are retained by pressing enter. In this exercise, we will use the CHANGE command to change the style of some previously drawn text. In Task 6, we will use CHANGE to alter some lines.

> Type "Change" or select the Point tool from the Resize flyout on the Modify toolbar, as shown in *Figure 7-14*.

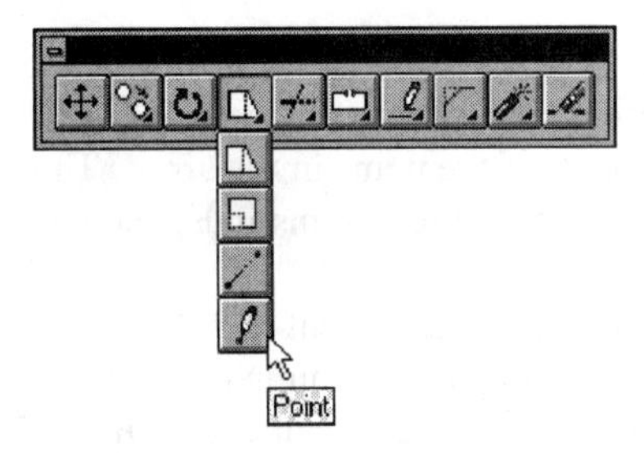

Figure 7-14

You will be prompted to select objects.

> Point to the word "Right".

> Press enter to end selection.

AutoCAD will issue the first of a series of prompts that ask you to specify what you would like to change:

Properties/<Change point>:

"Properties" refers to a standard list of options, including color, layer, and linetype which can also be changed using CHPROP, DDMODIFY, or DDCHPROP. CHPROP works exactly the same as CHANGE, except that it skips this prompt and moves directly to property change options. DDCHPROP provides the same options but in a dialogue box format.

"Change point" has different meanings with different types of entities. In the case of text, it allows you to relocate the text (this can also be done with the MOVE command or by moving a grip point).

> Press enter to skip property changes.

AutoCAD prompts:

Enter text insertion point:

This option allows you to move text in a drawing.
> Press enter to leave the text justification point where it is.
A prompt will be issued to allow you to change styles:

Text style: Standard
New style or RETURN for no change:

Let's change to our Vertical text style.
> Type "vertical".
You are now prompted for a change in rotation angle:

New rotation angle <270>:

> Press enter to retain vertical orientation.
Finally, AutoCAD prompts for a change in the characters themselves:

New Text <Right>:

This gives you yet another way to change wording and spelling. It is especially useful in a procedure where you create an array of one text item and then change the text of some or all of the items in the array. This is often more efficient than creating a large number of related text items independently. We will show you this technique in Drawing 7-2, "Gauges."
> Press enter to retain the current text.
This will terminate the command and your screen will be redrawn with "Right" in the Vertical text style, as shown in *Figure 7-15.*

NOTE: Grips can be used to edit text in the usual grip edit modes of moving, copying, rotating, mirroring, and scaling. The stretch mode works the same as moving. Grips on

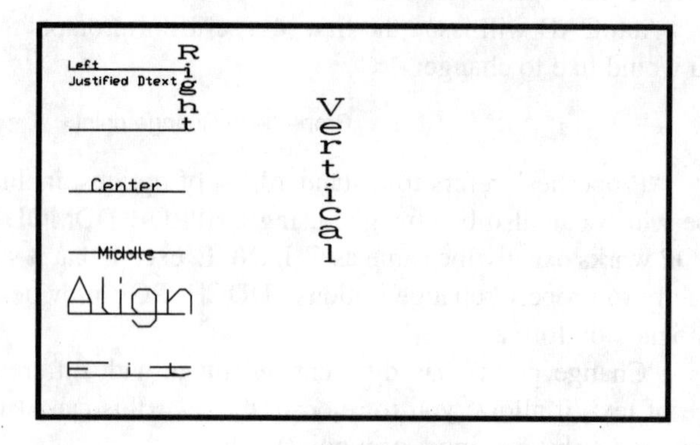

Figure 7-15

text are located at the left-justified start point and at the original point actually used to position the text. Grips cannot be used to reword, respell, or change text properties.

TASK 7: Using Change Points with Other Entities

You can use a change point to alter the size of lines and circles. With lines, CHANGE will perform a function similar to the EXTEND command, but without the necessity of defining an extension boundary. With circles, CHANGE will cause them to be redrawn so that they pass through the change point.

> Repeat the CHANGE command.

You will be prompted to:

Select objects:

> Select the first two lines, under the words "Left" and "Right".

> Press enter to end selection.

AutoCAD prompts:

Properties/<Change point>:

We will use a change point first to alter the selected lines, as shown in *Figure 7-16.*

> If ortho is on, turn it off (F8).

> Pick a point between the two lines and to the right, in the neighborhood of (6.50,10.00).

The lines will be redrawn so that the change point becomes the new end point of both lines.

To maintain horizontal or vertical orientation of lines while using the change point option of the CHANGE command, turn ortho on. Try it.

> Turn ortho on.

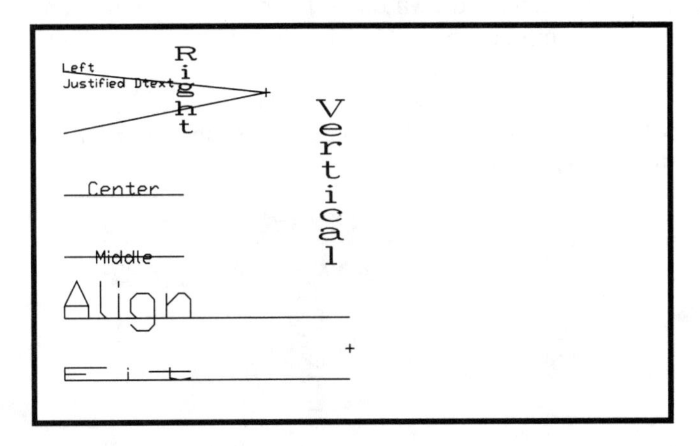

Figure 7-16

> Repeat the CHANGE command.
> Select the fifth and sixth lines where the words "Align" and "Fit" are drawn.
> Press enter to end selection.
> Pick a change point to the right of the lines, as shown by the blip in *Figure 7-16.*

Be careful not to pick this point too low or too high. This could cause AutoCAD to redraw one or more of the lines vertically. Both lines will be extended horizontally, as in *Figure 7-16.*

NOTE: If you use the change point option to edit a circle, the circle will be redrawn so that it passes through the change point. See *Figure 7-17.*

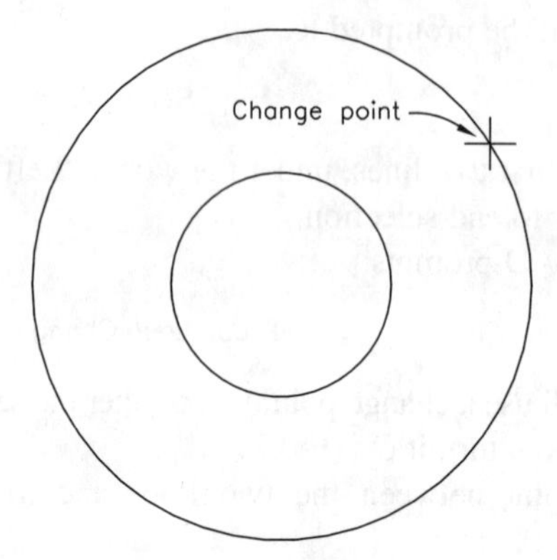

Figure 7-17

TASK 8: SCALEing Previously Drawn Entities

Any object or group of objects can be scaled up or down using the SCALE command or the grip edit scale mode. In this exercise, we will practice scaling some of the text and lines that you have drawn on your screen. Remember, however, that there is no special relationship between SCALE and text and that other types of entities can be scaled just as easily.

> Type "Scale" or select the Scale tool from the resize flyout on the Modify toolbar, as shown in *Figure 7-18.*

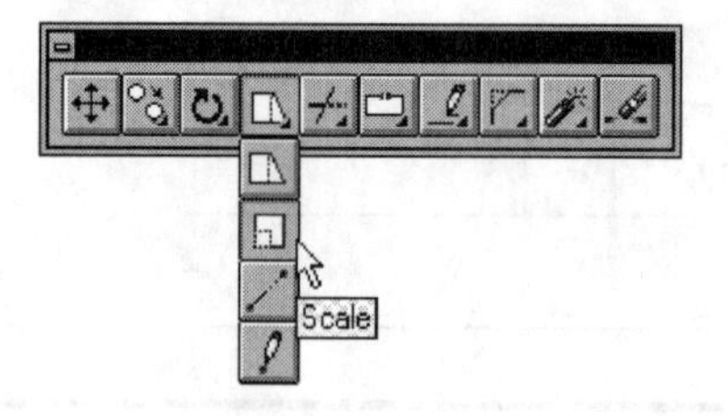

Figure 7-18

AutoCAD will prompt you to select objects.

> Use a crossing box (right to left) to select the set of six lines and text drawn in Task 1.
> Press enter to end selection.

You will be prompted to pick a base point:

Base point:

Imagine for a moment that you are looking at a square and you want to shrink it using a scale-down procedure. All the sides will, of course, be shrunk the same amount, but how do you want this to happen? Should the lower left corner stay in place and the whole square shrink toward it? Or should everything shrink toward the center? Or toward some other point on or off the square (see *Figure 7-19*)? This is what you

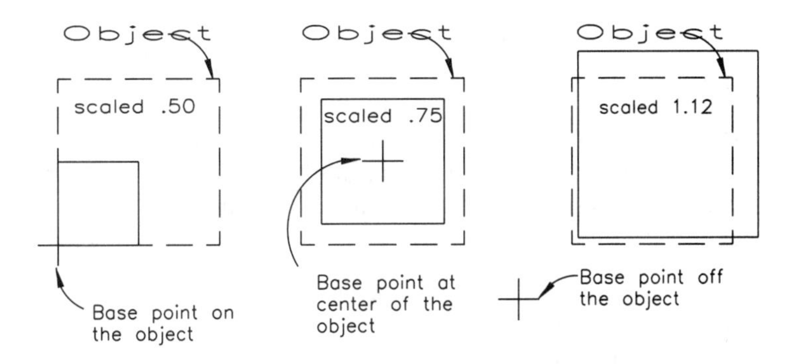

Figure 7-19

will tell AutoCAD when you pick a base point. In most applications you will choose a point somewhere on the object.
> Pick a base point at the left end of the bottom line of the selected set (the blip in *Figure 7-20*).

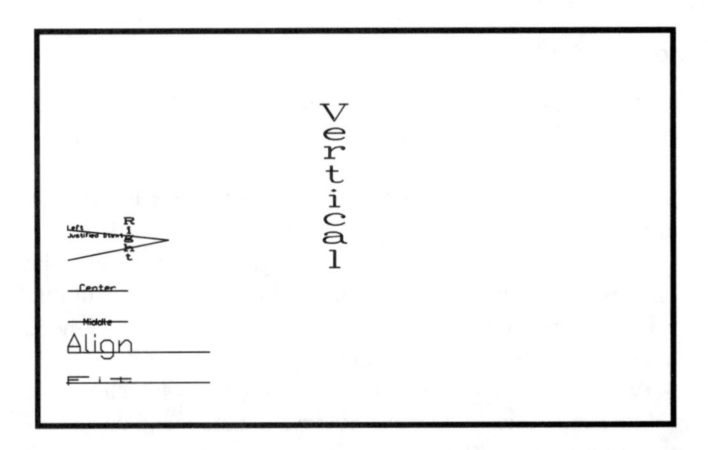

Figure 7-20

AutoCAD now needs to know how much to shrink or enlarge the objects you have selected:

<Scale factor>/Reference:

When you enter a scale factor, all lengths, heights, and diameters in your set will be multiplied by that factor and redrawn accordingly. Scale factors are based on a unit of 1. If you enter .5, objects will be reduced to half their original size. If you enter 2, objects will become twice as large.
> Type ".5" and press enter.

Your screen should now resemble *Figure 7-20.*

SCALEing by Reference

This option can save you from doing the arithmetic to figure out scale factors. It is useful when you have a given length and you know how large you want that length to become after the scaling is done. For example, we know that two of the lines we just scaled are now 2.00 long (the middle two that we did will not alter with CHANGE). Let's say that we want to scale them again to become 2.33 long (a scale factor of 1.165, but who wants to stop and figure that out?). This could be done using the following procedure:

1. Enter the SCALE command.
2. Select the "previous" set.
3. Pick a base point.
4. Type "r" or select "reference".
5. Type "2" for the reference length.
6. Type "2.33" for the new length.

NOTE: You can also perform reference scaling by pointing. In this procedure, you could point to the ends of the 2.00 line for the reference length and then show a 2.33 line for the new length.

Scaling with Grips

Scaling with grips is very similar to scaling with the SCALE command. To illustrate this, try using grips to return the text you just scaled back to its original size.

> Use a window or crossing box to select the six lines and the text drawn in Task 1 again.

There will be a large number of grips on the screen, three on each line and two on most of the text entities. Some of these will overlap or duplicate each other.
> Pick the grip at the lower left corner of the word "Fit," the same point used as a base point in the last scaling procedure.
> Type "sc" or press enter three times to bypass stretch, move, and rotate.
> Move the cursor slowly and observe the dragged image.

AutoCAD will use the selected grip point as the base point for scaling unless you specify that you want to pick a different base point.

Notice that you also have a reference option as in the SCALE command. Unlike the SCALE command, you also have an option to make copies of your objects at different scales.

As in SCALE, the default method is to specify a scale factor by pointing or typing.

> Type "2" or show a length of 2.00.

Your text will return to its original size and your screen will resemble *Figure 7-16* again.

> Press Esc twice to clear grips.

TASKS 9, 10, and 11

The drawings that follow contain typical applications of the text and modification commands you have just explored. Pay particular attention to the suggestions on the use of DTEXT in the first drawing, and CHANGE in the second and third drawings. These will save you time and help you to become a more efficient CAD operator.

DRAWING 7-1: TITLE BLOCK

This title block will give you practice in using a variety of text styles and sizes. You may want to save it and use it as a title block for future drawings.

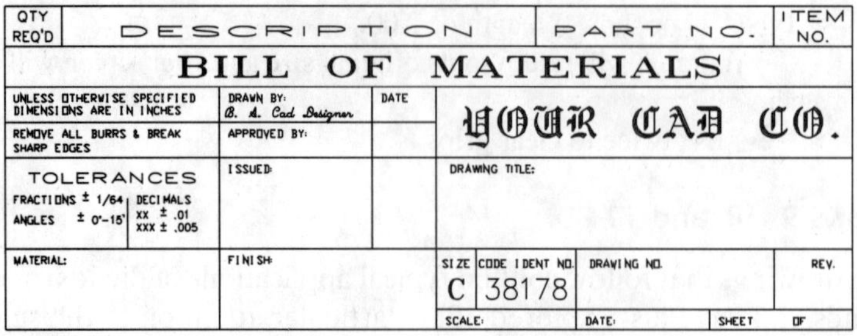

QTY REQ'D
DESCRIPTION
PART NO.
ITEM NO.
BILL OF MATERIALS
UNLESS OTHERWISE SPECIFIED DIMENSIONS ARE IN INCHES
DRAWN BY: B. A. Cad Designer
DATE
YOUR CAD CO.
REMOVE ALL BURRS & BREAK SHARP EDGES
APPROVED BY:
TOLERANCES
FRACTIONS ± 1/64
ANGLES ± 0°-15'
DECIMALS
XX ± .01
XXX ± .005
ISSUED:
DRAWING TITLE:
MATERIAL:
FINISH:
SIZE C
CODE IDENT NO. 38178
DRAWING NO.
REV.
SCALE:
DATE:
SHEET OF

DRAWING SUGGESTIONS

GRID = 1
SNAP = .0625

> Make ample use of DIST and TRIM as you draw the line patterns of the title block. Take your time and make sure that at least the major divisions are in place before you start entering text into the boxes.
> Set to the "text" layer before entering text.
> Use DTEXT with all the STANDARD, .09, left-justified text. This will allow you to do all of these in one command sequence, moving the cursor from one box to the next and entering the text as you go.
> Remember that once you have defined a style you can make it current using the TEXT or DTEXT commands. This will save you from having to restyle more than necessary.
> Type "%%D" to generate the degree symbol and "%%P" for the plus or minus symbol.

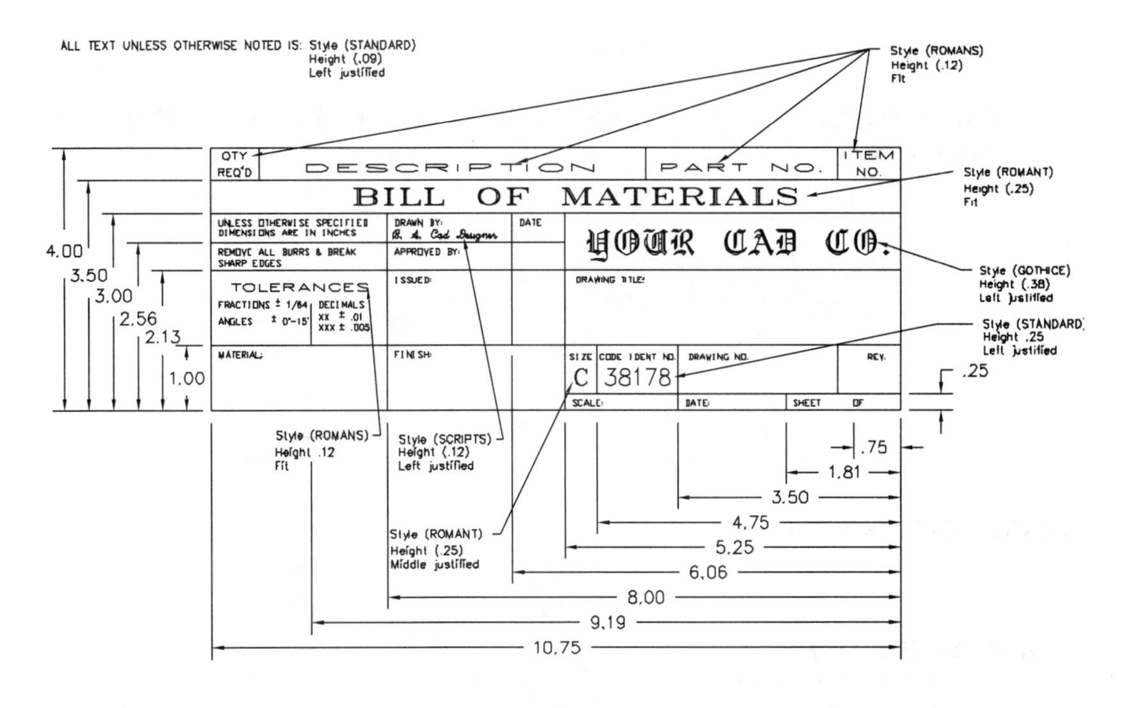

TITLE BLOCK

Drawing 7–1

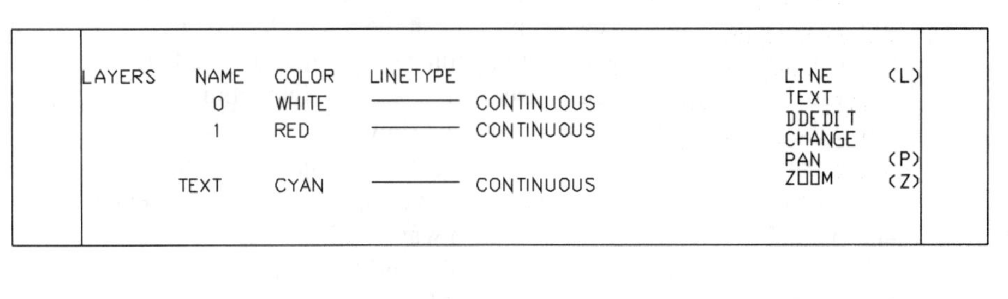

LAYERS	NAME	COLOR	LINETYPE	
	0	WHITE	————	CONTINUOUS
	1	RED	————	CONTINUOUS
	TEXT	CYAN	————	CONTINUOUS

LINE (L)
TEXT
DDEDIT
CHANGE
PAN (P)
ZOOM (Z)

F1	F2	F6	F7	F8	F9
HELP	TEXT/GRAPHICS SCREEN	ABSOLUTE/OFF/POLAR COORDS	ON/OFF GRID	ON/OFF ORTHO	ON/OFF SNAP

DRAWING 7-2: GAUGES

This drawing will teach you some typical uses of the SCALE and CHANGE commands. Some of the techniques used will not be obvious, so read the suggestions carefully.

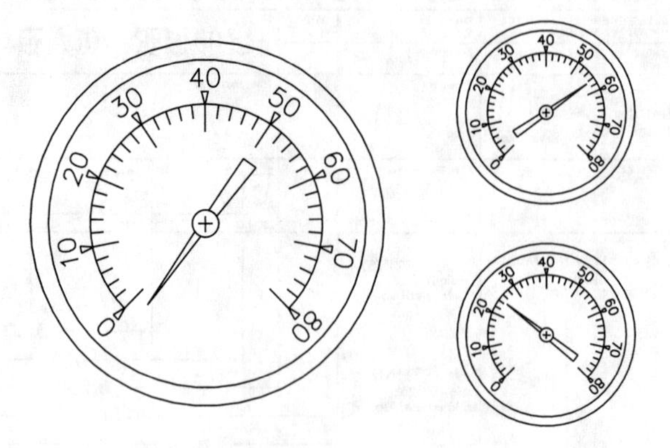

DRAWING SUGGESTIONS

GRID = .5
SNAP = .125

> Draw three concentric circles at diameters of 5.0, 4.5, and 3.0. The bottom of the 3.0 circle can be trimmed later.
> Zoom in to draw the arrow-shaped tick at the top of the 3.0 circle. Then draw the .50 vertical line directly below it and the number "0" (middle-justified text) above it.
> These three objects can be arrayed to the left and right around the perimeter of the 3.0 circle using angles of +135 and −135 as shown.
> Use the CHANGE command to change the arrayed zeros into 10, 20, 30, etc.
> Draw the .25 vertical tick directly on top of the .50 mark at top center and array it left and right. There will be 20 marks each way.
> Draw the needle horizontally across the middle of the dial.
> Make two copies of the dial; use SCALE to scale them down as shown. Then move them into their correct positions.
> Rotate all three needles as shown.

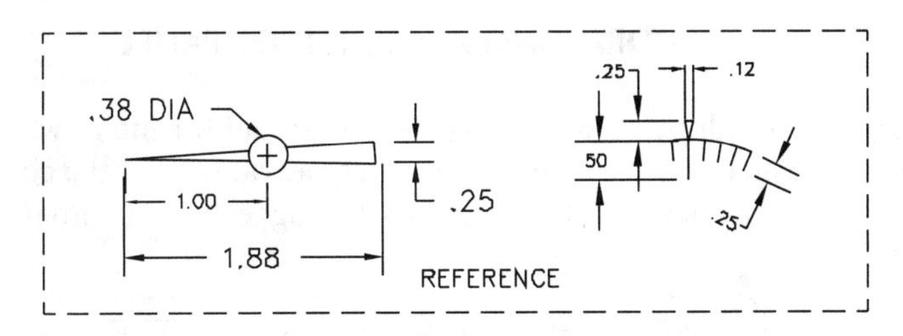

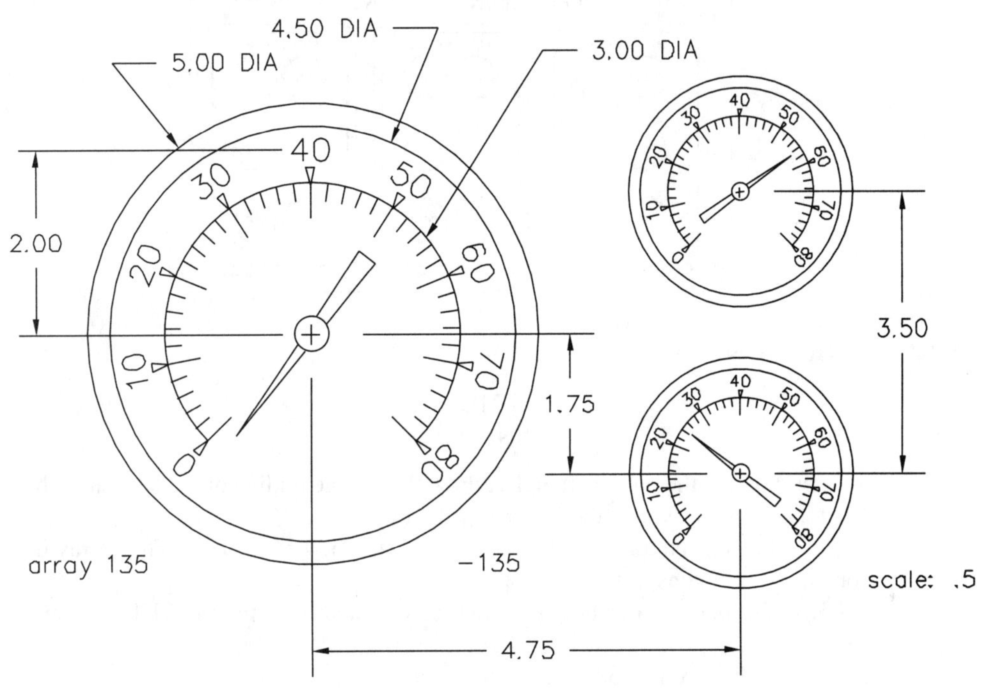

GAUGES
Drawing 7-2

LAYERS	NAME	COLOR	LINETYPE	
	0	WHITE	————	CONTINUOUS
	1	RED	————	CONTINUOUS
	3	GREEN	— — —	CENTER
	TEXT	CYAN	————	CONTINUOUS

LINE (L)
CIRCLE (C)
TEXT
SCALE
ROTATE
COPY
ARRAY
CHANGE
DDEDIT

F1	F2	F6	F7	F8	F9
HELP	TEXT/GRAPHICS SCREEN	ABSOLUTE/OFF/POLAR COORDS	ON/OFF GRID	ON/OFF ORTHO	ON/OFF SNAP

DRAWING 7-3: CONTROL PANEL

Done correctly, this drawing will give you a good feel for the power of the commands you now have available to you. Be sure to take advantage of the combinations of ARRAY and CHANGE described. Also, read the suggestion on moving the origin before you begin.

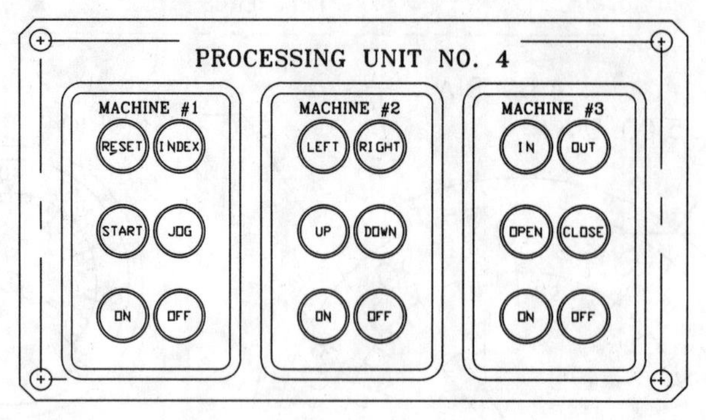

DRAWING SUGGESTIONS

GRID = .50
SNAP = .0625

> After drawing the outer rectangles, draw the double outline of the left button box and fillet the corners. Notice the different fillet radii.

> Draw the "on" button with its text at the bottom left of the box. Then array it 2 × 3 for the other buttons in the box.

> CHANGE the lower right button text to "off" and draw the MACHINE # text at the top of the box.

> ARRAY the box 1 × 3 to create the other boxes.

> CHANGE the text for the buttons and machine numbers as shown.

> Complete the drawing.

MOVING THE ORIGIN WITH THE UCS COMMAND

The dimensions of this drawing are shown in ordinate form, measured from a single point of origin in the lower left-hand corner. In effect, this establishes a new coordinate origin. If we move our origin to match this point, then we will be able to read dimension values directly from the coordinate display. This may be done by setting the lower left-hand limits to (−1,−1). It also may be done using the UCS command to establish a User Coordinate System with the origin at a point you specify. User Coordinate Systems are discussed in Chapter 9. For now, here is a simple procedure:

> Type "ucs".

> Type "o" for the "Origin" option.

> Point to the new origin.

That's all there is to it. Move your cursor to the new origin and watch the coordinate display. It should show "0.00,0.00", and all values will be measured from there.

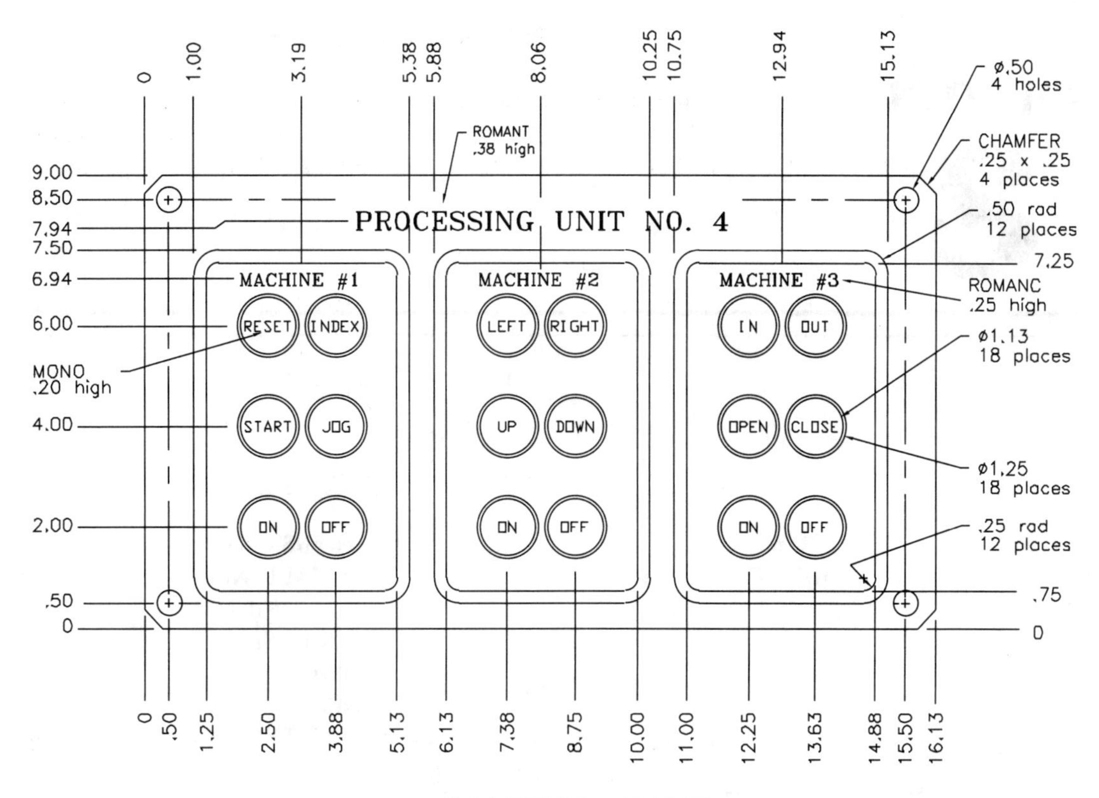

CONTROL PANEL

Drawing 7-3

LAYERS	NAME	COLOR	LINETYPE	
	0	WHITE	———	CONTINUOUS
	1	RED	———	CONTINUOUS
	3	GREEN	— — —	CENTER
	TEXT	CYAN	———	CONTINUOUS

LINE (L)
CIRCLE (C)
TEXT
COPY
ARRAY
FILLET
CHAMFER
DDEDIT

F1	F2	F6	F7	F8	F9
HELP	TEXT/GRAPHICS SCREEN	ABSOLUTE/OFF/POLAR COORDS	ON/OFF GRID	ON/OFF ORTHO	ON/OFF SNAP

CHAPTER 8

COMMANDS

DRAW	DATA	DIMENSIONING
BHATCH	DIM	DIMALIGNED
		DIMANGULAR
		DIMBASELINE
		DIMCONTINUE
		DIMLINEAR
		DIMSTYLE

OVERVIEW

The ability to dimension your drawings and add crosshatch patterns will greatly enhance the professional appearance and utility of your work. AutoCAD's dimensioning feature is a complex system of commands, subcommands, and variables which automatically measure objects and draw dimension text and extension lines. With AutoCAD's dimensioning tools and variables, you can create dimensions in a wide variety of formats, and these formats can be saved as styles. The time saved through not having to draw each dimension line by line is very significant.

TASKS

1. Define and save a dimension style.
2. Draw linear dimensions (horizontal, vertical, and aligned).
3. Draw multiple linear dimensions (baseline and continue).
4. Draw angular dimensions.
5. Draw center marks and diameter and radius dimensions.
6. Add cross hatching to previously drawn objects.

7. Do Drawing 8-1 ("Flanged Wheel").
8. Do Drawing 8-2 ("Shower Head").
9. Do Drawing 8-3 ("Plot Plan").

TASK 1: Creating and Saving a Dimension Style

Dimensioning in AutoCAD is highly automated and very easy compared to manual dimensioning. In order to achieve a high degree of automation while still allowing for the broad range of flexibility required to cover all dimension styles, the AutoCAD dimensioning system is necessarily quite complex. In the exercises that follow we will guide you through the system, show you some of what is available, and give you a good foundation for understanding how to get what you want out of AutoCAD dimensioning. We will create a basic dimension style and use it to draw standard dimensions. We will leave it to you to explore the many variations that are possible.

In Release 13 it is best to begin by defining a dimension style. A dimension style is a set of dimension variable settings that control the text and geometry of all types of AutoCAD dimensions. In previous releases, the default styles for most dimensions would parallel the drawing units set through the Units Control dialogue box or the UNITS command. Release 13 dimensions, however, are quite independent of drawing units and must be set separately. We recommend that you create the new dimension style in your prototype drawing and save it. Then you will not have to make these changes again when you start new drawings.

> To begin this exercise, type "open" or select the Open tool from the Standard toolbar.

This calls up the Select File dialogue box.

> In the File Name box, type "b" or select whatever drawing you are using as a prototype.

When "B" is highlighted you should see the preview image in the black box at the right. This will be blank if your prototype drawing has no objects drawn in it.

> Click on "OK".

This will open your prototype drawing. Now we will make some changes in dimension style settings so that all dimensions showing distances will be presented with two decimal places and angular dimensions will have no decimals.

> Type "ddim" or select "Dimension Style . . ." from the Data pull down menu.

This will call up the Dimension Styles dialogue box shown in *Figure 8-1*. You will see that the current dimension style is called "Standard". Among other things, the AutoCAD Standard dimension style uses four-place decimals in all dimensions, including angular.

On the left below the Dimension Style name box, you will see the Family box with a set of radio buttons for the different types of dimensions. A family is a complete group of dimension settings for the different types of dimensions. The types of dimensions are shown with radio buttons: Linear, Radial, Angular, etc. You can set these individually or all at once to a single format using the Parent radio button. In this exercise, we will first use the Parent button to set all types of dimensions to two-place decimals. Then we will set angular dimensions individually to show no decimal places.

First, let's give the new dimension style a name. We will use "B" to go along with our B prototype.

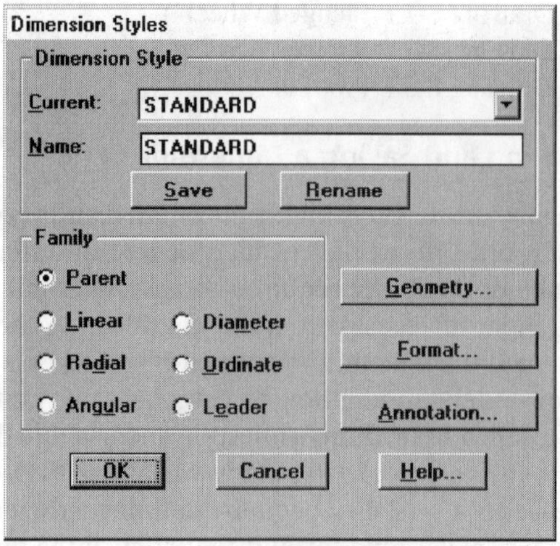

Figure 8-1

> Double click in the Name: edit box to highlight the word "Standard".
> Type "B".
> Click on "Save".

This will create the new dimension style and make it current, so that subsequent dimensions will be drawn using the B style.

> Check to make sure that the Parent radio button is selected. It should be by default.

Before moving on, look at the three call boxes at the right. We will make changes only in the Annotation ... subdialogue, but you may wish to look at the others while you are here. If you do, simply open the dialogue boxes and cancel them when you are through looking. Geometry ... calls up a subdialogue that allows you to adjust aspects of the lines, arrowheads, extension lines, and center marks that make up a dimension. Format ... allows for changes in the placement of dimension text relative to dimension geometry. Annotation ... allows for changes in the text and measurements provided for AutoCAD dimensions.

> Click on "Annotation ...".

This will call up the Annotation dialogue box shown in *Figure 8-2.* There are areas here for adjusting Primary Units (which is what we are interested in), Tolerance, Alternate Units, and Text style. We will explore tolerances later. Alternate units are dimension units that may be automatically presented in parentheses along with primary units. Text styles were discussed in the last chapter. Standard will be the default text style for dimensions, but if you have another text style defined in your drawing, you could select it here. Then all dimensions drawn with this dimension style would have the selected text style.

> Click on "Units ..." in the Primary Units box on the upper left of the dialogue box.

This will bring up the Primary Units dialogue box shown in *Figure 8-3.* There are adjustments available for units, dimension precision, angular units, and tolerances. The lists under Units and Angles are the same lists used by the UNITS command. The Units

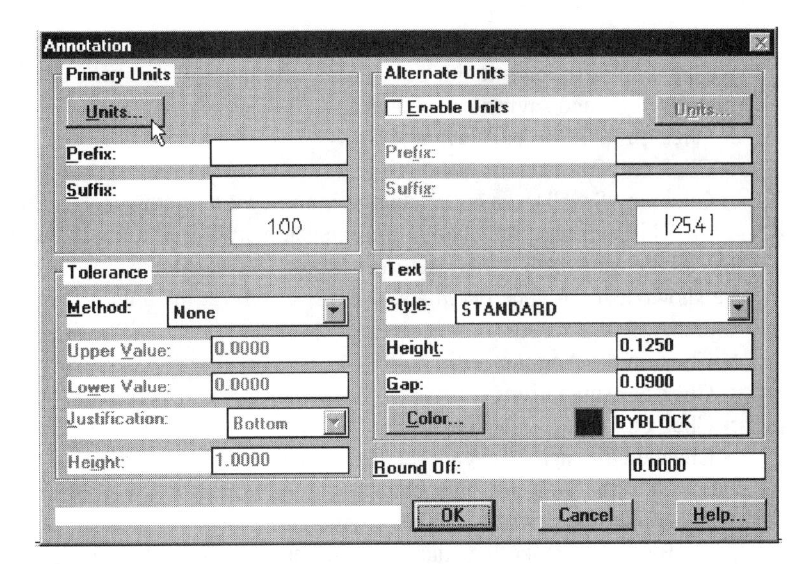

Figure 8-2

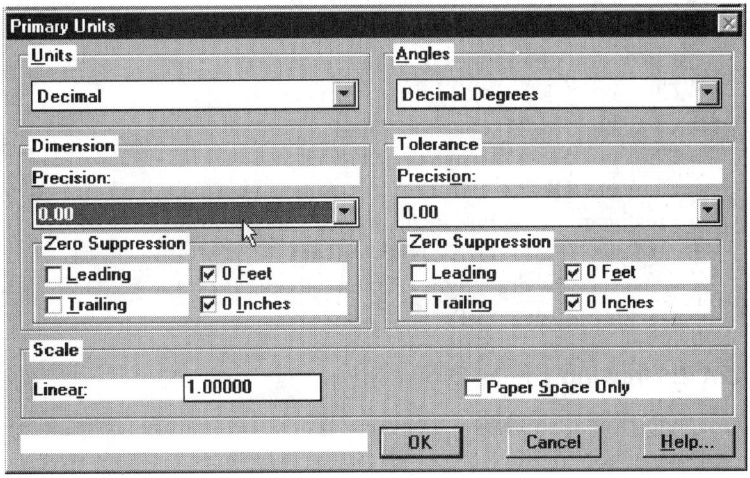

Figure 8-3

should show "Decimal" and the Angles box should show "Decimal Degrees". If for any reason these are not showing in your box, you should make these changes now. For our purposes, all we need to change is the number of decimal places showing in the Dimension Precision box. By default it will be 0.0000. We will change it to 0.00.

> Click on the arrow to the right of the Precision box.

This will open a list of precision settings ranging from 0 to 0.00000000.

> Click on 0.00.

This will close the list box and show 0.00 as the selected precision. At this point, we are ready to complete this part of the procedure by returning to the Dimension Style dialogue box and saving our changes.

> Click on "OK" to exit Primary Units.

> Click on "OK" to exit Annotation.

> Click on "Save" to save the change in dimension precision.

This will save the change in dimension style B. At this point dimension style B has all the AutoCAD default settings except that all dimensions, linear and angular, will be shown as two-place decimals. Next, we will set the angular precision separately to 0 places.

> Click on the Angular radio button.

> Click on Annotation . . ..

> Click on Units . . ..

> Change Dimension Precision to 0 decimal places.

Note that you are only changing angular dimension precision, all other dimensions in the family will keep the parent setting of 0.00.

Be sure to make the change in the same Precision box that you used before. It is the one on the left, under Dimension, not the one on the right under Tolerance. The two boxes are identical, and since the Tolerance box is directly under the Angle box it may seem natural to make the change there. The Tolerance box only changes the precision of tolerances, which we are not using right now. The Dimension box is used to adjust the precision of all types of dimensions.

> Click on "OK" to exit Primary Units.

> Click on "OK" to exit Annotation.

> Click on "Save" to save the changes to B.

> Click on "OK" to exit the Dimension Styles dialogue box.

> Finally, go to the File menu, save prototype drawing B with the new dimension style B, and use New to open a new drawing.

If the new drawing is opened with the B prototype, the B dimension style will be current for the next task.

TASK 2: Drawing Linear Dimensions

AutoCAD Release 13 has many commands and features that aid in the drawing of dimensions. Prior to Release 13 all AutoCAD dimensioning was handled through the DIM command, which was really a subsystem of commands for different types of dimensions. In Release 13 the DIM subcommands have become commands that can be accessed directly from the command prompt.

In this exercise, you will create some basic linear dimensions in the now current B style.

> To prepare for this exercise, draw a triangle (ours is 3.00, 4.00, 5.00) and a 6.00 line, as shown in *Figure 8-4*. Exact sizes and locations are not critical.

We will begin by adding dimensions to the triangle.

The new Release 13 dimensioning commands are streamlined and efficient. Their full names, however, are all rather long. They all begin with "dim" and are fol-

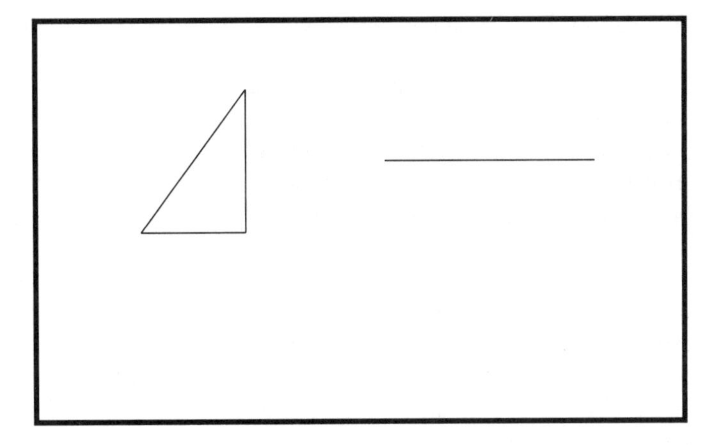

Figure 8-4

lowed by what used to be the DIM command options (for example: DIMLINEAR, DIMALIGNED, and DIMANGULAR). They do have somewhat shorter aliases, which we will show you as we go along. We encourage you to use the Dimensioning toolbar to avoid typing these names.

We begin by placing a linear dimension below the base of the triangle.

> To open the Dimensioning toolbar, select "Toolbars" and then "Dimensioning" from the Tools pull down menu.

This will open the Dimensioning toolbar shown in *Figure 8-5*.

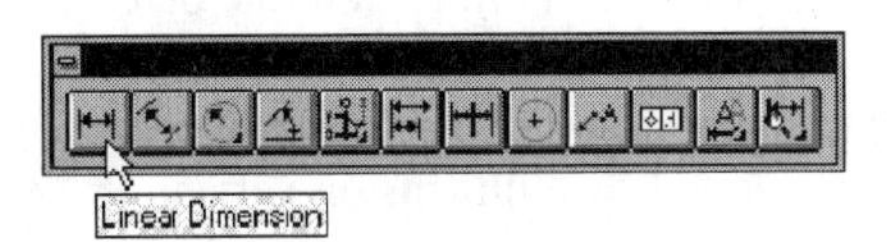

Figure 8-5

> Type "dimlin" or select the Linear Dimension tool from the Dimensioning toolbar, as shown.

All of these methods will bring you into the DIMLINEAR command, with the following prompt in the command area:

First extension line origin or RETURN to select:

There are two ways to proceed at this point. One is to show where the extension lines should begin, and the other is to select the base of the triangle itself and let AutoCAD position the extension lines. In most cases, the select method is faster.

> Press enter (RETURN) to indicate that you will select an object.

AutoCAD will replace the cross hairs with a pickbox and prompt for your selection:

Select object to dimension:

> Select the horizontal line at the bottom of the triangle, as shown by point 1 in *Figure 8-6.*

AutoCAD immediately creates a dimension, including extension lines, dimension line, and text, that you can drag away from the selected line. AutoCAD will place the dimension line and text where you indicate, but will keep the text centered between the extension lines. The prompt is as follows:

Dimension line location (Text/Angle/Horizontal/Vertical/Rotated):

In the default sequence, you will simply show the location of the dimension. If you wish to alter the text, you can do so by using the "Text" option or change it later with a command called DIMEDIT. "Angle", "Horizontal", and "Vertical" allow you to specify the orientation of the text. Horizontal text is the default for linear text. "Rotated" allows you to rotate the complete dimension, so that the extension lines move out at an angle from the object being dimensioned (text remains horizontal).

NOTE: If the dimension variable "dimsho" is set to 0 (off), you will not be given an image of the dimension to drag into place. The default setting is 1 (on), so this should not be a problem. If, however, it has been changed in your drawing, type "dimsho" and then "1" to turn it on again.

> Pick a location about .50 below the triangle, as shown by point 2 in *Figure 8-6.*

Bravo! You have completed your first dimension.

Notice that our figure and others in this chapter are shown zoomed in on the relevant object for the sake of clarity. You may zoom or not as you like.

At this point, take a good look at the dimension you have just drawn to see what it consists of. As in *Figure 8-6,* you should see the following components: two extension lines, two "arrows," a dimension line on each side of the text, and the text itself.

Notice also that AutoCAD has automatically placed the extension line origins a short distance away from the triangle base (you may need to zoom in to see this). This distance is controlled by a dimension variable called "dimexo", which can be changed

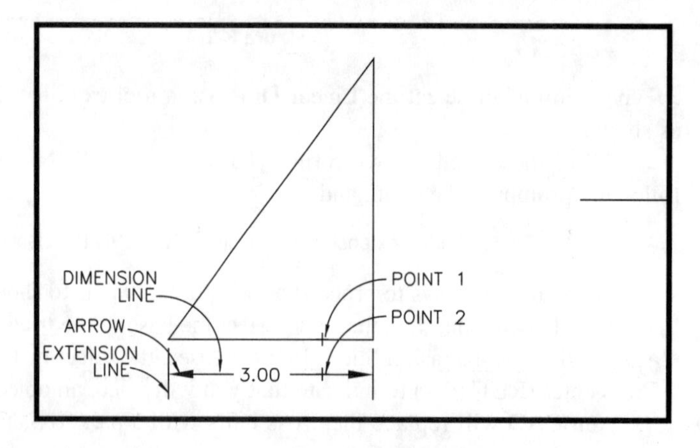

Figure 8-6

in the Dimension Style dialogue box under Geometry. It is one of many variables that control the look of AutoCAD dimensions. Another example of a dimension variable is "dimasz", which controls the size of the arrows at the end of the extension lines.

Next, we will place a vertical dimension on the right side of the triangle. You will see that DIMLINEAR handles both horizontal and vertical dimensions.

> Repeat the DIMLINEAR command.

You will be prompted for extension line origins as before:

First extension line origin or RETURN to select:

This time we will show the extension line origins manually.

> Pick the right angle corner at the lower right of the triangle, point 1 in *Figure 8-7*. AutoCAD will prompt for a second point:

Second extension line origin:

Even though you are manually specifying extension line origins, it is not necessary to show the exact point where you want the line to start. AutoCAD will automatically set the dimension lines slightly away from the line as before, according to the setting of the dimexo dimension variable.

> Pick the top intersection of the triangle, point 2 in the figure.

From here on, the procedure will be the same as before. You should have a dimension to drag into place and the following prompt:

Dimension line location (Text/Angle/Horizontal/Vertical/Rotated):

Your screen should now include the vertical dimension, as shown in *Figure 8-7*.

Now let's place a dimension on the diagonal side of the triangle. For this, we will need the DIMALIGNED command.

> Type "dimali" or select the Aligned Dimension tool from the Dimensioning toolbar, as shown in *Figure 8-8*.

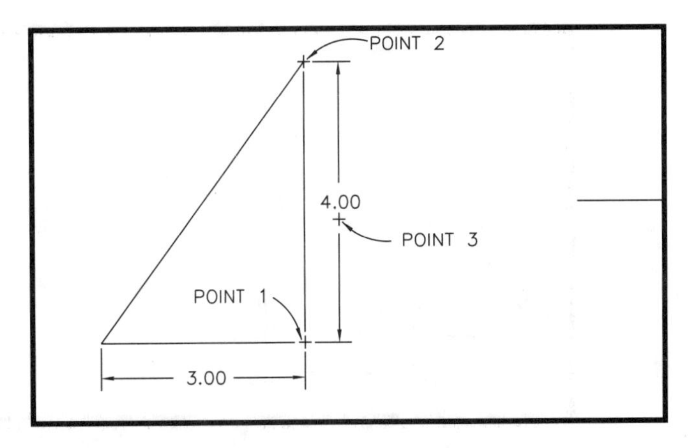

Figure 8-7

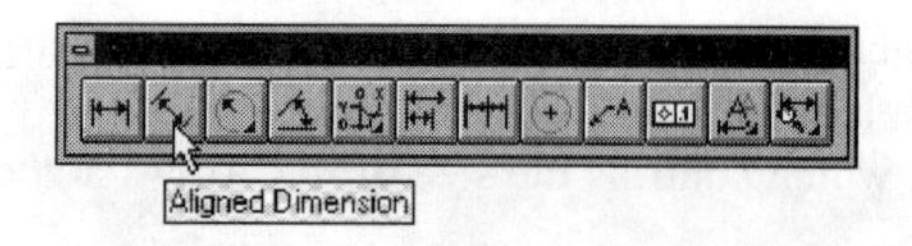

Figure 8-8

> Press enter (RETURN), indicating that you will select an object.

AutoCAD will give you the pickbox and prompt you to select an object to dimension.

> Select the hypotenuse of the triangle.

> Pick a point approximately .50 above and to the left of the line.

Your screen should resemble *Figure 8-9.* Notice that AutoCAD retains horizontal text in aligned and vertical dimensions as the default.

TASK 3: Drawing Multiple Linear Dimensions—Baseline and Continued

DIMBASELINE and DIMCONTINUE allow you to draw multiple linear dimensions efficiently. In baseline format you will have a series of dimensions all measured from the same initial origin. In continued dimensions there will be a string of dimensions in which the second extension line for one dimension becomes the first extension for the next.

> To prepare for this exercise, be sure that you have a 6.00 horizontal line as shown in *Figure 8-4* at the beginning of Task 2. Although the figures in this exercise will show only the line, leave the triangle in your drawing because we will come back to it in the next task.

In this exercise, we will be placing a set of baseline dimensions on top of the line and a continued series on the bottom. In order to use either Baseline or Continue, you

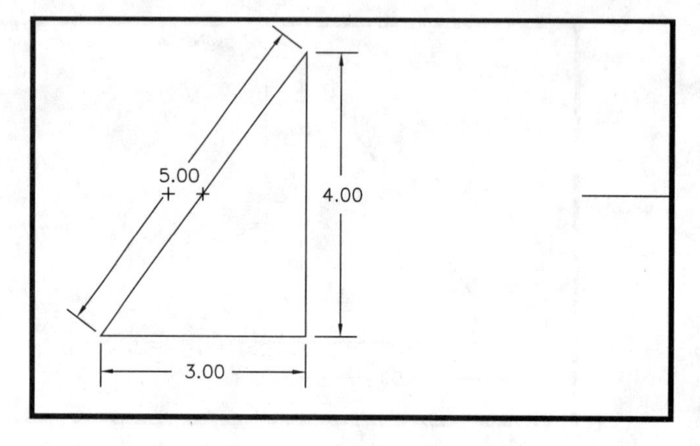

Figure 8-9

must have one linear dimension already drawn on the line you wish to dimension. So we will begin with this.

> Select the Linear Dimension tool from the Dimensioning toolbar.
> Pick the left end point of the line for the origin of the first extension line.
> Pick a second extension origin 2.00 to the right of the first, as shown in *Figure 8-10.*
> Pick a point .50 above the line for dimension text.

You should now have the initial 2.00 dimension shown in *Figure 8-10.* We will

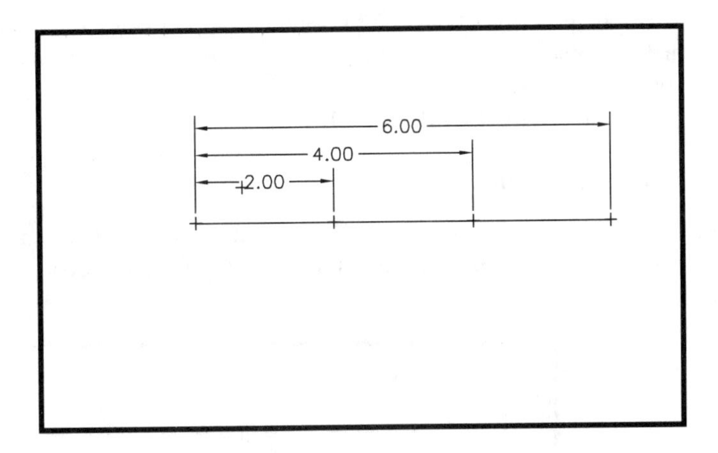

Figure 8-10

use DIMBASELINE to add the other dimensions above it.

> Type "dimbase" or select the Baseline Dimension tool from the Dimensioning toolbar, as shown in *Figure 8-11.*

AutoCAD uses the first extension line origin you picked again and prompts for a second:

```
Second extension line origin or RETURN to select:
```

> Pick a point 4.00 to the right of the first extension line (in other words, 4.00 from the left end of the 6.00 line).

The second baseline dimension shown in *Figure 8-10* should be added to your drawing. We will add one more. AutoCAD leaves you in the DIMBASELINE command so that you can add as many baseline dimensions as you want.

> Pick the right end point of the line.

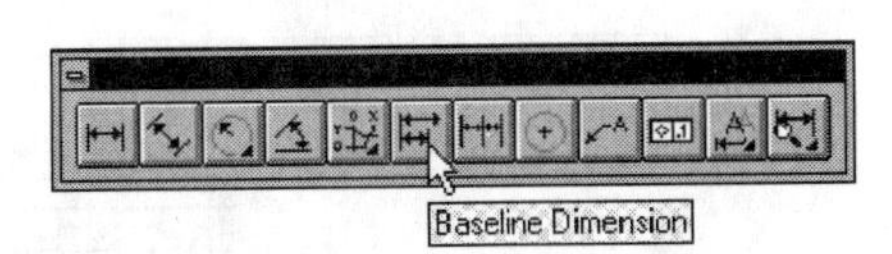

Figure 8-11

Your screen should resemble *Figure 8-10.* See how quickly AutoCAD draws dimensions? As long as you have your dimension style defined the way you want it, this can make dimensioning very easy.

AutoCAD will still be prompting for another second extension line origin. We will need to press enter twice to exit DIMBASELINE.

> Press enter to end extension line selection.

The first time you press enter, AutoCAD will show you a new prompt:

Select base dimension:

This prompt would allow you to begin a new series of baseline dimensions from a dimension other than the last one you drew. Bypassing this prompt will take you back to command prompt.

> Press enter to exit DIMBASELINE.

Continued Dimensions

Now we will place three continued dimensions along the bottom of the line, as shown in *Figure 8-12.* This process is very similar to the baseline process so you should need little help at this point.

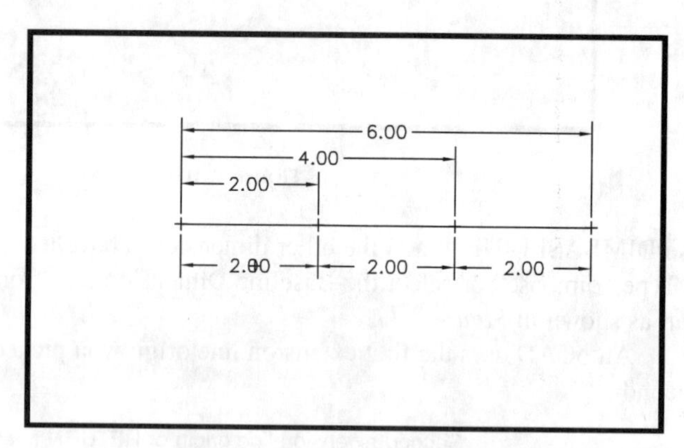

Figure 8-12

> Begin by using DIMLINEAR to place an initial horizontal dimension .50 below the line, showing a length of 2.00 from the left end, as shown in *Figure 8-12.*

> Type "dimcont" or select the Continue Dimension tool from the Dimensioning toolbar, as shown in *Figure 8-13.*

> Pick a second extension line origin 2.00 to the right of the last extension line.

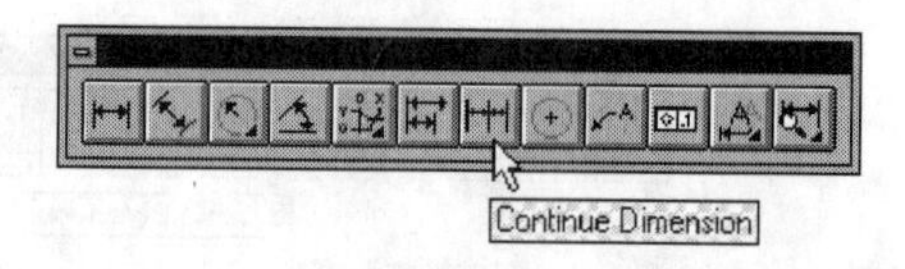

Figure 8-13

Be sure to pick this point on the 6.00 line; otherwise, the extension line will be left hanging.

> Pick another second extension line at the right end of the line.

> Press enter to end the continued series.

> Press enter to exit DIMCONTINUE.

Done!

TASK 4: Drawing Angular Dimensions

Angular dimensioning works much like linear dimensioning, except that you will be prompted to select objects that form an angle. AutoCAD will compute an angle based on the geometry that you select (two lines, an arc, part of a circle, or a vertex and two points) and construct extension lines, a dimension arc, and text specifying the angle.

For this exercise, we will return to the triangle and add angular dimensions to two of the angles as shown in *Figure 8-14.*

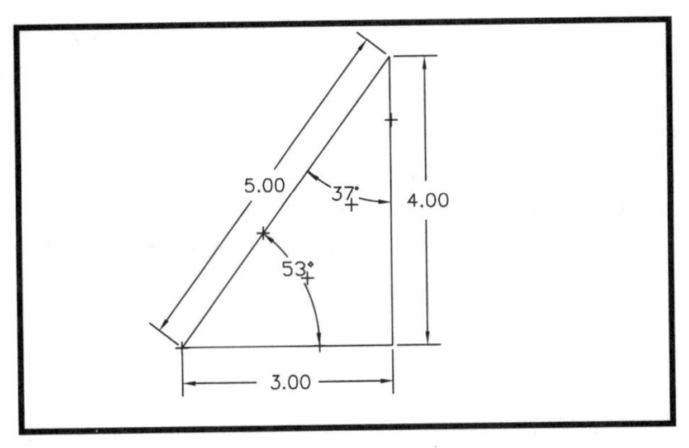

Figure 8-14

> Type "dimang" or select the Angular Dimension tool from the Dimensioning toolbar, as shown in *Figure 8-15.*

The first prompt will be:

```
Select arc, circle, line, or RETURN:
```

The prompt shows that you can use angular dimensions to specify angles formed by arcs and portions of circles as well as angles formed by lines. If you press enter (RETURN), you can specify an angle manually by picking its vertex and a point on each side of the angle. We will begin by picking lines, the most common use of angular dimensions.

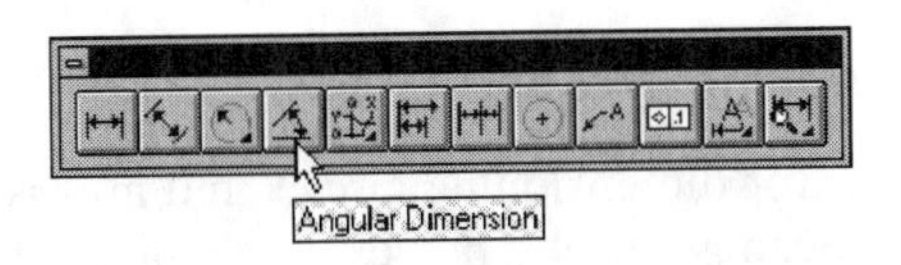

Figure 8-15

> Select the base of the triangle.

You will be prompted for another line:

Second line:

> Select the hypotenuse.

As in linear dimensioning, AutoCAD now shows you the dimension lines and lets you drag them into place (assuming dimsho is on). The prompt asks for a dimension arc location and also allows you the option of changing the text or the text angle.

Dimension arc line location (Text/Angle):

> Move the cursor around to see how the dimension may be placed and then pick a point between the two selected lines, as shown by the blip in *Figure 8-14*.

The lower left angle of your triangle should now be dimensioned, as in *Figure 8-14*. Notice that the degree symbol is added by default in angular dimension text.

We will dimension the upper angle by showing its vertex, using the RETURN option. This time we will place the text outside the arc, as shown in *Figure 8-14*.

> Repeat DIMANGULAR.
> Press enter (RETURN).

AutoCAD prompts for an angle vertex.

> Point to the vertex of the angle at the top of the triangle.

AutoCAD prompts:

First angle endpoint:

> Pick a point along the hypotenuse.

In order to be precise this should be a snap point. The most dependable one will be the lower left corner of the triangle.

AutoCAD prompts:

Second angle endpoint:

> Pick any point along the vertical side of the triangle.

There should be many snap points on the vertical line, so you should have no problem.

> Move the cursor slowly up and down within the triangle.

Notice how AutoCAD places the arrows outside the angle when you approach the vertex and things get crowded. Also notice that if you move outside the angle, AutoCAD switches to the outer angle.

> Pick a location for the dimension arc as shown in *Figure 8-14*.

TASK 5: Dimensioning Arcs and Circles

The basic process for dimensioning circles and arcs is as simple as those we have already covered. It can get tricky, however, when AutoCAD does not place the dimension where you want it. Text placement can be controlled by adjusting dimension variables. In this exercise, we will create a center mark and some diameter and radius dimensions.

> To prepare for this exercise, draw two circles as shown in *Figure 8-16.* The circles we have used have radii of 2.00, 1.50.

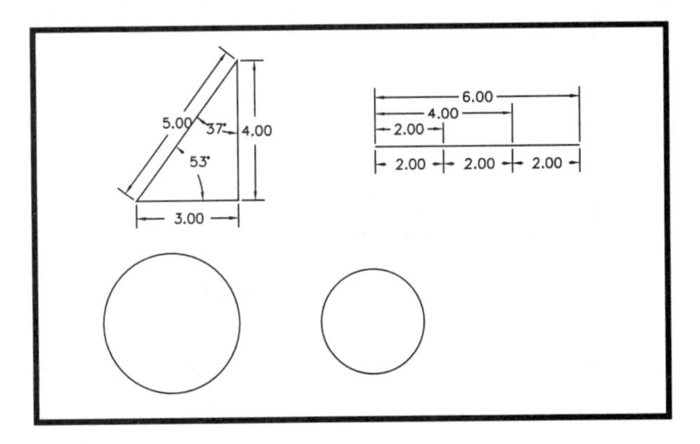

Figure 8-16

> Type "dimcenter" or select the Center Mark tool from the Dimensioning toolbar, as shown in *Figure 8-17.*

You cannot use "dimcen" as the command name because that is the name of the dimension variable that controls the style of the center mark.

Center marks resemble blips, but they are actual lines and will appear on a plotted drawing. They are the simplest of all dimension features to create and are created automatically as part of some radius and diameter dimensions.

AutoCAD prompts:

```
Select arc or circle:
```

> Select the smaller circle.

A center mark will be drawn in the 1.50 circle, as shown in *Figure 8-18.* You may want to REDRAW your display to see the difference between blips and center marks. Blips are in white and, if you are drawing on layer 1, your center mark will be red.

NOTE: A different type of center mark can be produced by changing the dimension variable "dimcen" from .09 to -.09. The result is shown in the dimension variables chart (*Figure 8-21*) at the end of this task.

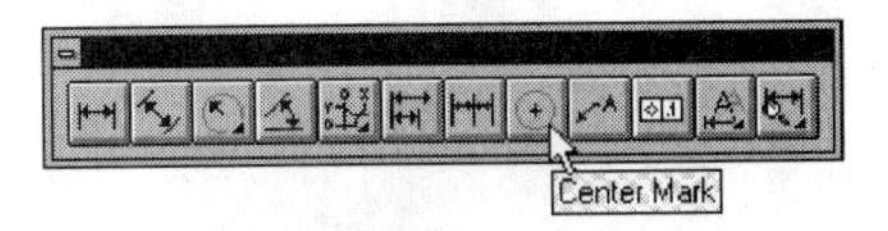

Figure 8-17

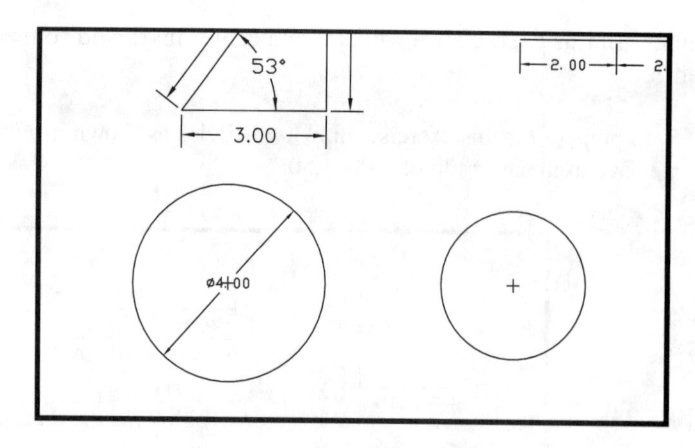

Figure 8-18

Now we will add the diameter dimension shown on the larger circle in *Figure 8-18.*

> Type "dimdia" or select the Diameter Dimension tool from the Radial dimension flyout on the Dimensioning toolbar, shown in *Figure 8-19.*

AutoCAD will prompt:

Select arc or circle:

> Select the larger circle.

AutoCAD will show a diameter dimension with the dimension text at the center of the circle and ask for Dimension line location. The Text and Angle options allow you to change the dimension text or put it at an angle. If you move your cursor around, you will see that you can position the dimension line anywhere around the circle.

> Pick a dimension position so that your screen resembles *Figure 8-20.*

Notice that the diameter symbol prefix is added automatically by default.

The placement of diameter dimensions is an important consideration. Diameter dimensions are often shown as this one is, at the center of the circle. This can be problematic if other dimensions, text, or objects are in the way. Release 13 offers substantial flexibility through the grip editing system.

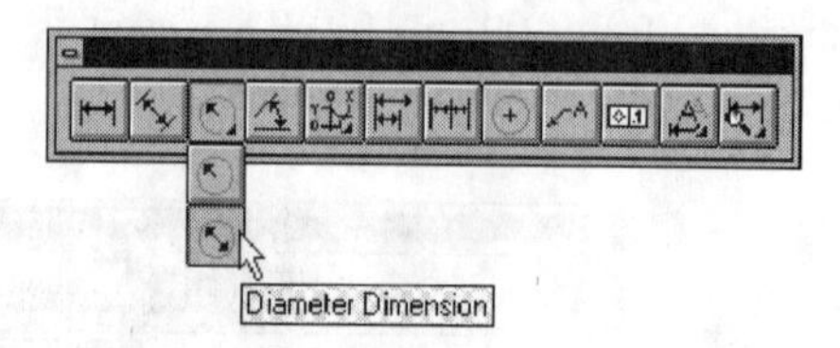

Figure 8-19

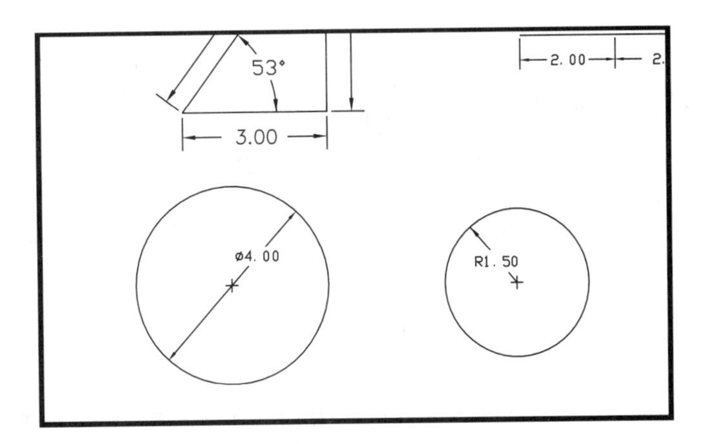

Figure 8-20

Using Grips to Move Dimension Text

Suppose you want to move the dimension text 4.00 away from the center of the circle, but keep it within the extension lines. This is easily done using grips.

> Pick the diameter dimension just drawn.

Dimensions are block entities and can be selected by picking any part of the block. Blocks are collections of objects treated as one entity.

You will know when you have picked the dimension by looking for the three blue grips boxes. You will find that the grip stretch mode works just like the move mode when you are working with a block.

> Pick the box at the center near the dimension text.

After picking the grip box, you will find that you can move both the dimension line and the text around inside the circle. Actually, you can move outside the circle as well. If you do this, AutoCAD will place the text outside and draw a leader to it from the dimension line.

> Pick a point above and to the right of the center to produce the text placement shown in *Figure 8-20.* Ortho must be off to do this.

Radius Dimensions

The procedures for radius dimensioning are exactly the same as those for diameter dimensions. Try adding a radius dimension to the smaller circle, as shown in *Figure 8-20.*

Dimension Variables

The chart in *Figure 8-21* shows some of the more common dimension variables that can be used to change the look and style of dimensions. Dimension variables can be changed through the Dimension Style dialogue box, or by typing the name of the variable at the Command prompt and then entering a new value.

COMMONLY USED DIMENSION VARIABLES					
VARIABLE	DEFAULT VALUE	APPEARANCE	DESCRIPTION	NEW VALUE	APPEARANCE
dimaso	on	All parts of dim are one entity	Associative dimensioning	off	All parts of dim are separate entities
dimscale	1.00	2.00	Changes size of text & arrows, not value	2.00	2.00
dimasz	.18		Sets arrow size	.38	
dimcen	.09		Center mark size and appearance	-.09	
dimdli	.38		Spacing between continued dimension lines	.50	
dimexe	.18		Extension above dimension line	.25	
dimexo	0.06		Extension line origin offset	.12	
dimtp	0.00	1.50	Sets plus tolerance	.01	$1.50^{+0.01}_{-0.00}$
dimtm	0.00	1.50	Sets minus tolerance	.02	$1.50^{+0.00}_{-0.02}$
dimtol	off	1.50	Generate dimension tolerances (dimtp & dimtm must be set) (dimtol & dimlim cannot both be on)	on	$1.50^{+0.01}_{-0.02}$
dimlim	off	1.50	Generate dimension limits (dimtp & dimtm must be set) (dimtol & dimlim cannot both be on)	on	1.51 1.48
dimtad	off	1.50	Places text above the dimension line	on	1.50
dimtxt	.18	1.50	Sets height of text	.38	1.50
dimtsz	.18	1.50	Sets tick marks & tick height	.25	1.50
dimtih	on	1.50	Sets angle of text When off rotates text to the angle of the dimension	off	1.50
dimtix	off	Ø0.71	Forces the text to inside of circles and arcs. Linear and angular dimensions are placed inside if there is sufficient room	on	Ø0.71

Figure 8-21

TASK 6: Using the BHATCH Command

Automated hatching is another immense timesaver. Release 13 includes two commands for use in cross hatching. Of the two, BHATCH and HATCH, BHATCH is more powerful and easier to use. BHATCH differs from HATCH in that it automatically defines the nearest boundary surrounding a point you have specified.

> To prepare for this exercise, clear your screen of all previously drawn objects and then draw three rectangles, one inside the other, with the word "TEXT" at the center, as shown in *Figure 8-22.*

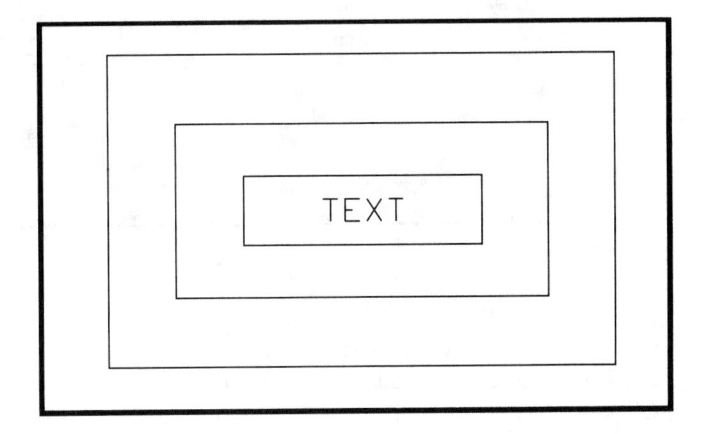

Figure 8-22

> Type "bhatch" or select the Hatch tool from the Draw toolbar, as shown in *Figure 8-23.*

Hatch on the toolbar actually enters the BHATCH command. Whether typed or selected, BHATCH always calls up the Boundary Hatch dialogue box shown in *Figure 8-24.* At the top of the box you will see the Pattern Type box. Before we can hatch anything, we need to specify a pattern. Later, we will show AutoCAD what we wish to hatch using the Pick Points method.

The Pattern type currently shown in the image box is ANSI131, a predefined pattern. There are about 68 of these patterns, but first we will use a simple User Defined pattern of straight lines on a 45 degree angle.

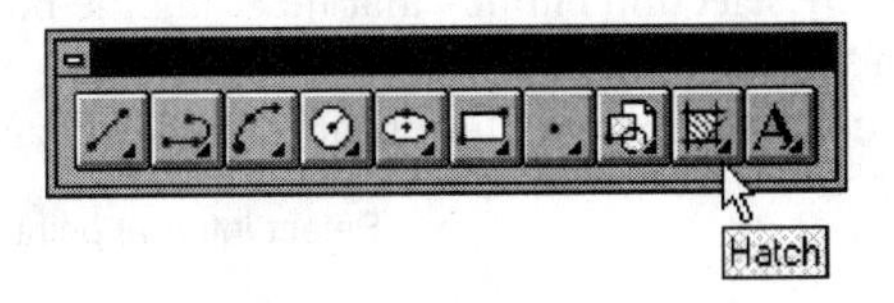

Figure 8-23

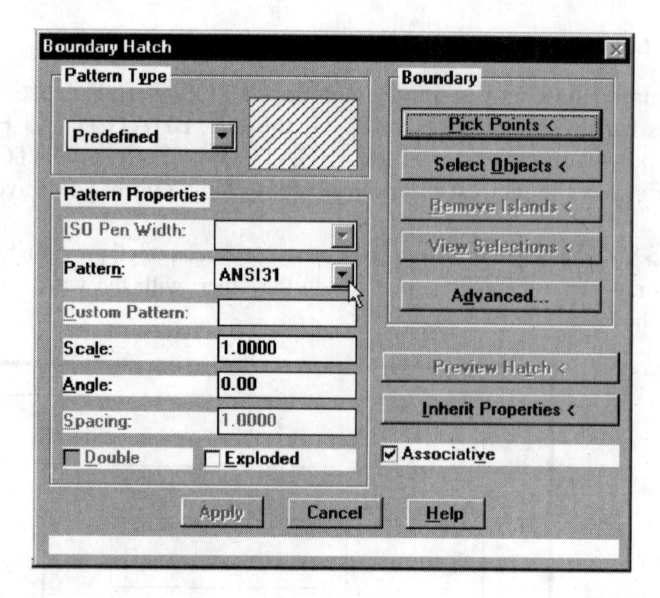

Figure 8-24

> Click on the arrow to the right of "Predefined".

This opens a list including Predefined, User-Defined, and Custom patterns.

> Click on "User-Defined".

When you select a user-defined pattern there will be nothing shown in the image box.

When you create a user pattern you will need to specify an angle and a spacing.

> Double click in the Angle edit box, and then type "45".

> Double click in the Spacing edit box, and then type ".5".

The remaining boxes in this area are Double and Exploded. Double hatching creates double hatch lines running perpendicular to each other according to the specified angle. Hatch patterns are created as a single entity within the specified boundary. If they are created with Exploded checked, each line of the pattern will be a separate entity.

Next we need to show AutoCAD where to place the hatching. To the right in the dialogue box is the Boundary area. Looking down the dialogue box, the next item is the "Pick Points" option. With this option, you can have AutoCAD locate a boundary when you point to the area inside it. The "Select Objects" option can be used to create boundaries by selecting entities that lie along the boundaries.

> Click on "Pick Points <".

The dialogue box will disappear temporarily and you will be prompted as follows:

```
Select internal point:
```

> Pick any point inside the largest outer rectangle, but outside the inner rectangles.

AutoCAD displays these messages, though you may have to press F2 to see them:

Selecting everything visible . . .
Analyzing the selected data . . .
Analyzing internal islands . . .

In a large drawing this process can be time consuming, as the program searches all visible entities to locate the appropriate boundary. When the process is complete all of the rectangles and text will be highlighted. It will happen very quickly in this case.

AutoCAD continues to prompt for internal points so that you can define multiple boundaries. Let's return to the dialogue box and see what we've done so far.

> Press enter to end internal point selection.

The dialogue box will reappear. You may not have noticed, but several of the options that were "grayed out" before are now accessible. We will make use of the "Preview Hatch" option. This allows us to preview what has been specified without leaving the command, so that we can continue to adjust the hatching until we are satisfied.

> Click on "Preview Hatch <".

Your screen should resemble *Figure 8-25,* except that you will have to move the Boundary Hatch - Continue box away from the center of the display. This demonstrates the "normal" style of hatching in which multiple boundaries are hatched or left clear in alternating fashion, beginning with the outermost boundary and working inward. You can see the effect of the other styles by looking at the Advanced Options dialogue box. Notice that BHATCH has recognized the text as well as the other interior boundaries.

> Click on "Continue" to return to the Boundary Hatch dialogue box.

You could complete the hatching operation at this point by clicking on Apply, but let's take a look at a few more details while we are here.

> Click on "Advanced . . .".

This calls up the Advanced Options dialogue box shown in *Figure 8-26.* In the middle you will see the hatching style area. There are three basic options that determine how BHATCH will treat interior boundaries. "Normal" is the current style.

> Click on the arrow to the right of "Normal".

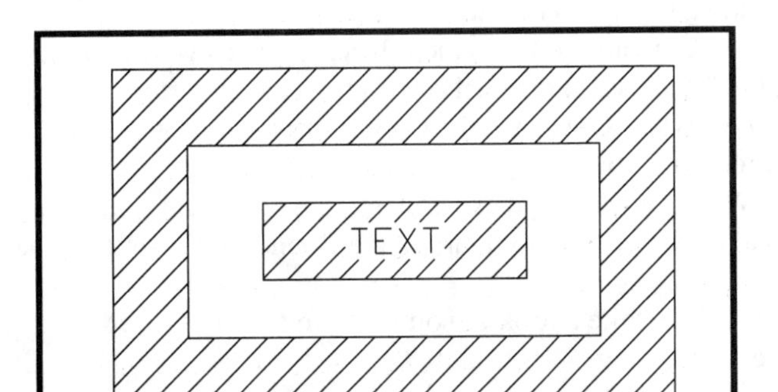

Figure 8-25

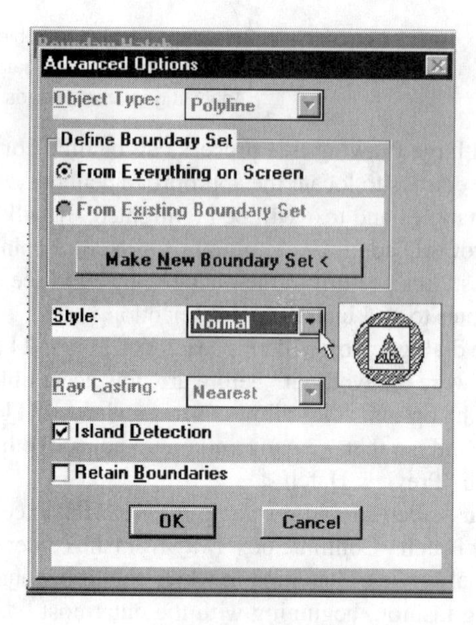

Figure 8-26

You will see the other two styles listed. "Normal" hatches alternate areas moving inward, "Outer" hatches only the outer area, and "Ignore" hatches through all interior boundaries. There is an image box that shows an example of the current style, as shown in *Figure 8-26.*

> Click on "Outer" and watch the black image box.

It now shows the "Outer" style in which only the area between the outermost boundary and the first inner boundary is hatched.

> Click on the arrow again and then on "Ignore" and watch the image box.

In the "Ignore" style, all interior boundaries are ignored and the entire area including the text is hatched.

You can see the same effects in your own drawing, if you like, by clicking on "OK" and then "Preview Hatch" after changing from the "Normal" style to either of the other styles. Of course, the three styles are indistinguishable if you do not have boundaries within boundaries to hatch.

> When you are done experimenting, you should select "Normal" style hatching again and click on "OK" to return to the Boundary Hatch dialogue box.

Now let's take a look at some of the fancier stored hatch patterns that AutoCAD provides.

> Click on the arrow to the right of "User-Defined" and select "Predefined".

> Now click on the arrow to the right of "Pattern" in the Pattern box.

This opens a long list of AutoCAD's predefined hatch patterns. To produce the hatched image in *Figure 8-27,* we chose the Escher pattern.

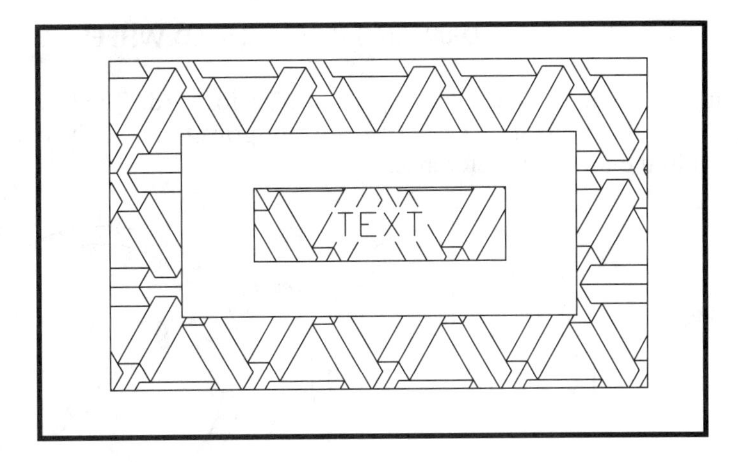

Figure 8-27

> Use the scroll bar to move down the list until you come to "Escher".
> Select "Escher".

As soon as the pattern is selected, you will see it illustrated in the pattern type image box.

To produce the figure, we also used a larger scale in this hatching.

> Double click in the "Scale:" edit box and type "1.5".
> Double click in the "Angle:" edit box and type "0".
> Click on "Preview Hatch <".

Your screen should resemble *Figure 8-27,* but remember this is still just the preview. When you are through adjusting hatching, you have the choice of canceling the BHATCH command so that no hatching is added to your drawing, or completing the process by clicking on "Apply". Apply will confirm the most recent choices of boundaries, patterns, and scales, making them part of your drawing.

> Click on "Continue".
> Click on "Apply" to exit BHATCH and confirm the hatching.

TASKS 7, 8, and 9

Dimensioning and Hatching are two of AutoCAD's most powerful features. They will do vast amounts of work for you and perform very well if used correctly. In general, you will complete other drawing procedures first and save hatching and dimensioning until the end. Dimensions and hatch patterns should also be allotted separate layers of their own so that they can be turned on and off at will.

Both dimensioning and hatching can be time consuming and require careful attention. Remember, in most applications your drawings will be of little use until the dimensions and appropriate hatching are clearly and correctly placed.

DRAWING 8-1: FLANGED WHEEL

Most of the objects in this drawing are straightforward. The keyway is easily done using the TRIM command. Use DIMEDIT or grips to move the diameter dimension as shown in the following reference.

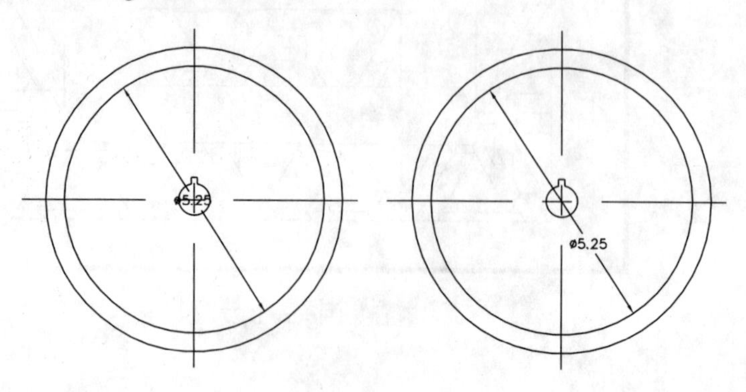

DRAWING SUGGESTIONS

GRID = .25
SNAP = .0625
HATCH line spacing = .50

> You will need a .0625 snap to draw the keyway. Draw a .125 × .125 square at the top of the .63 diameter circle. Drop the vertical lines down into the circle so that they may be used to TRIM the circle. TRIM the circle and the vertical lines, using a window to select both as cutting edges.

> Remember to set to layer "hatch" before hatching, layer "text" before adding text, and layer "dim" before dimensioning.

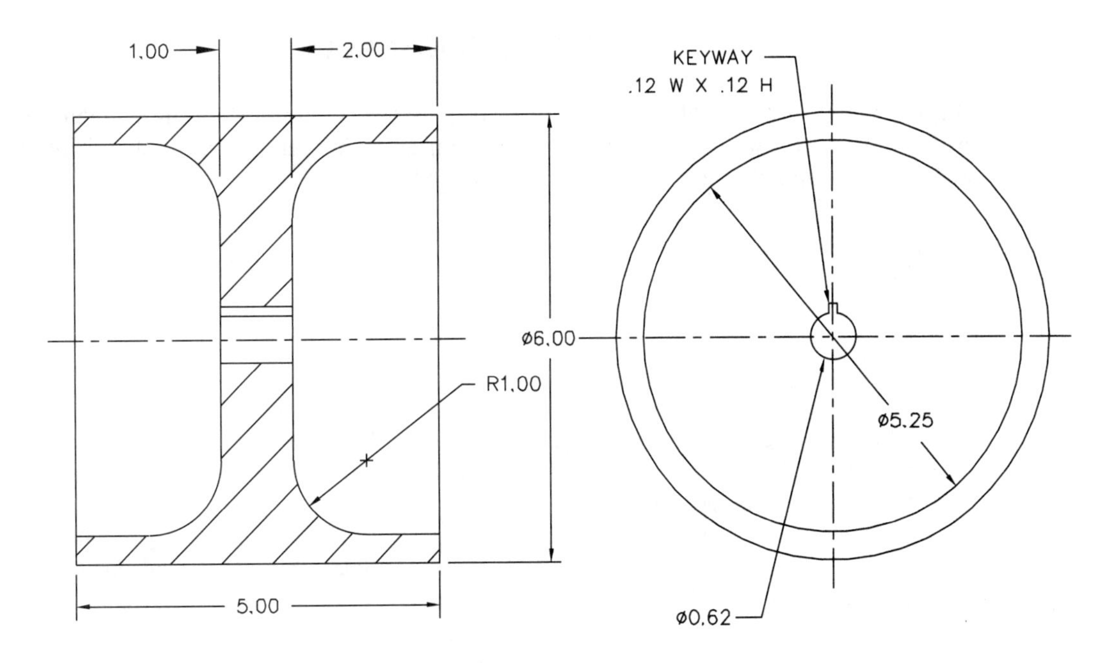

FLANGED WHEEL

Drawing 8-1

LAYERS	NAME	COLOR	LINETYPE
	0	WHITE	CONTINUOUS
	1	RED	CONTINUOUS
	3	GREEN	CENTER
	TEXT	CYAN	CONTINUOUS
	HATCH	BLUE	CONTINUOUS
	DIM	MAGENTA	CONTINUOUS

LINE (L)
CIRCLE (C)
MOVE (M)
BREAK
HATCH
FILLET
DIM
ZOOM (Z)

F1	F2	F6	F7	F8	F9
HELP	TEXT/GRAPHICS SCREEN	ABSOLUTE/OFF/POLAR COORDS	ON/OFF GRID	ON/OFF ORTHO	ON/OFF SNAP

DRAWING 8-2: SHOWER HEAD

This drawing makes use of the procedures for hatching and dimensioning you learned in the last two drawings. In addition, it uses an angular dimension, baseline dimensions, leaders, and "%%c" for the diameter symbol.

DRAWING SUGGESTIONS

GRID = .50
SNAP = .125
HATCH line spacing = .25

> You can save some time on this drawing by using MIRROR to create half of the right side view. Notice, however, that you cannot hatch before mirroring, because the mirror command will reverse the angle of the hatch lines.
> To achieve the angular dimension at the bottom of the right side view, you will need to draw the vertical line coming down on the right. Select this line and the angular line at the right end of the shower head, and the angular extension will be drawn automatically. Add the text "2 PL" using the DTEXT command.
> Notice that the diameter symbols in the vertical dimensions at each end of the right side view are not automatic. Use "%%c" to add the diameter symbol to the text.

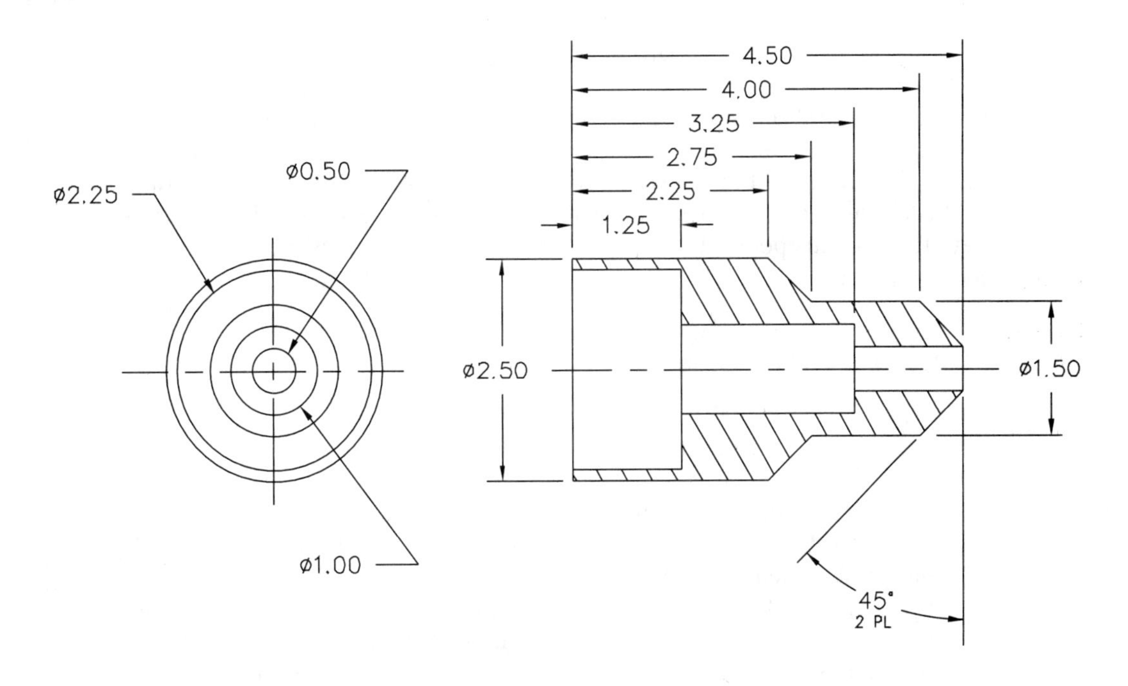

SHOWER HEAD

Drawing 8-2

LAYERS	NAME	COLOR	LINETYPE	
	0	WHITE	———	CONTINUOUS
	1	RED	———	CONTINUOUS
	2	YELLOW	– – – –	HIDDEN
	3	GREEN	—– - —–	CENTER
	TEXT	CYAN	———	CONTINUOUS
	HATCH	BLUE	———	CONTINUOUS
	DIM	MAGENTA	———	CONTINUOUS

LINE (L)
CIRCLE (C)
MIRROR
HATCH
DIM
ZOOM (Z)

F1	F2	F6	F7	F8	F9
HELP	TEXT/GRAPHICS SCREEN	ABSOLUTE/OFF/POLAR COORDS	ON/OFF GRID	ON/OFF ORTHO	ON/OFF SNAP

DRAWING 8-3: PLOT PLAN

This architectural drawing makes use of three hatch patterns and several dimension variable changes. Be sure to make these settings as shown. Notice that we have simplified the format of the drawing page for this drawing. This is because the drawings are becoming more involved and because you should need less information to complete them at this point. We will continue to show drawings this way for the remainder of the book.

DRAWING SUGGESTIONS

GRID = 10′	LIMITS = 180′,120′
SNAP = 1′	LTSCALE = 2′

> The "trees" shown here are symbols for oaks, willows, and evergreens.

> Use the DIST command to find start points for the inner rectangular objects (the garage, the dwelling, etc.).

> BHATCH will open a space around text inside a defined boundary; however, sometimes you will want more white space than BHATCH leaves. The simple solution is to draw a box around the text area as an inner boundary. If the BHATCH style is set to "Normal", it will stop hatching at the inner boundary. Later you can erase the box, leaving an island of white space around the text.

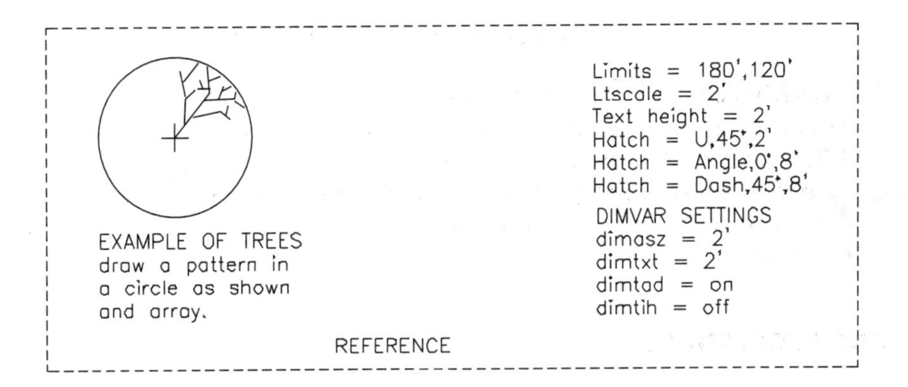

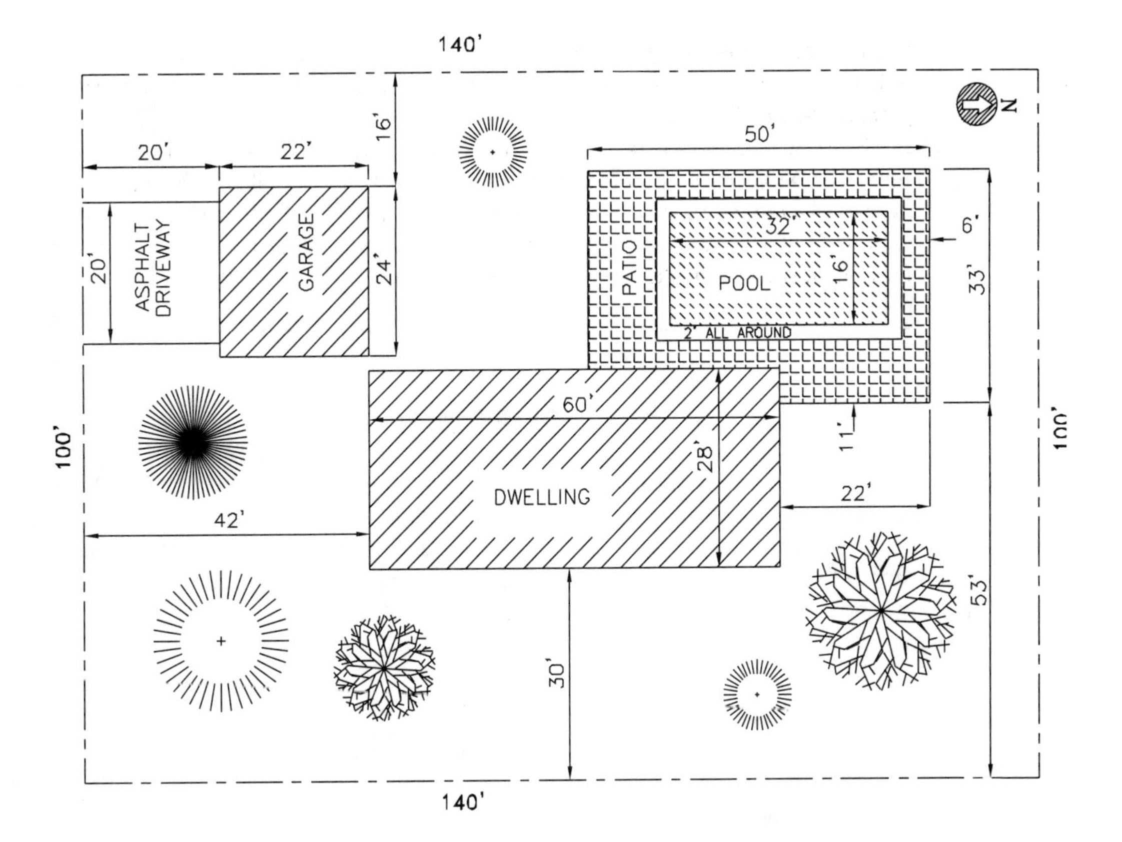

PLOT PLAN

Drawing 8–3

CHAPTER 9

COMMANDS

VIEW	OPTIONS	SURFACES
UCS	UCSICON	3DFACE
VPOINT		EDGESURF
DDVPOINT	**SOLIDS**	REVSURF
	BOX	RULESURF
	WEDGE	TABSURF

OVERVIEW

It is now time to begin thinking in three dimensions. 3D drawing in AutoCAD is logical and efficient. You can create wireframe models, surface models, or solid models and display them from multiple points of view. In this chapter we will focus on User Coordinate Systems, 3D viewpoints, and the three types of 3D models. These are the primary tools you will need to understand how AutoCAD allows you to work in three dimensions on a two-dimensional screen.

TASKS

1. Create and view a 3D box.
2. Define and save three User Coordinate Systems.
3. Use standard edit commands in a UCS.
4. Use multiple tiled viewports.
5. Create flat surfaces with 3DFACE.
6. Remove hidden lines with HIDE.
7. Use 3D polygon mesh commands.

8. Create solid boxes and wedges.
9. Create the union of two solids.
10. Working above the XY plane with ELEV.
11. Create the subtraction of two solids.
12. SHADE and RENDER solid objects.
13. Do Drawing 9-1 ("Clamp").
14. Do Drawing 9-2 ("REVSURF Designs").
15. Do Drawing 9-3 ("Bushing Mount").

TASK 1: Creating and Viewing a 3D Wireframe Box

In this task, we will create a simple 3D box that we can edit in later tasks to form a more complex object.

In Chapter 1 we showed how to turn the coordinate system icon (see *Figure 9-1*) off and on. For drawing in 3D you will definitely want it on. If your icon is not visible, follow this procedure to turn it on:

1. Type "Ucsicon" or select "UCS" from the Options pull down menu.
2. Type "on" or select "Icon".

For now, simply observe the icon as you go through the process of creating the box, and be aware that you are currently working in the same coordinate system that you have always used in AutoCAD. It is called the World Coordinate System, to distinguish it from others you will create yourself beginning in Task 2.

Currently, the origin of the WCS is at the lower left of your screen. This is the point (0,0,0) when you are thinking 3D, or simply (0,0) when you are in 2D. x coordinates increase to the right horizontally across the screen, and y coordinates increase vertically up the screen as usual. The Z axis, which we have ignored up until now, currently extends out of the screen towards you and perpendicular to the X and Y axes. This orientation of the three planes is also called a plan view. Soon we will switch to a "front, right, top" or "southeast" view.

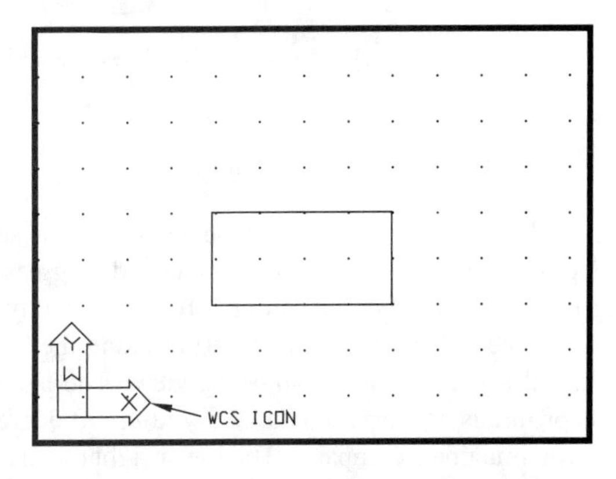

Figure 9-1

Let's begin.

> Draw a 2.00 by 4.00 rectangle near the middle of your screen, as shown in *Figure 9-1*. Do not use the RECTANG command to draw this figure because we will want to select individual line segments later on.

Changing Viewpoints

In order to move immediately into a 3D mode of drawing and thinking, our first step will be to change our viewpoint on this object.

In Release 13 the simplest and most efficient method is to use the 3D Viewpoints Presets submenu from the View pull down menu. For our purposes, this is the only method you will need.

> Select "View" from the pull down menu bar.
> Select "3D Viewpoint Presets" from the View menu.

This will call up a cascading submenu beginning with "Plan View", as shown in *Figure 9-2*. Using this method, AutoCAD will take you directly into one of 11 preset

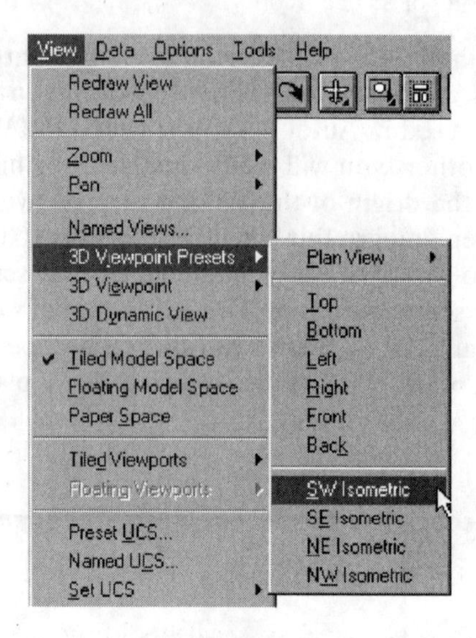

Figure 9-2

viewpoints. In all of the non-isometric views in the submenu (Plan View, Top, . . ., Front, Back), the X and Y axes will be rotated 90 degrees or not at all. You will see the drawing from directly above (Plan and Top) or directly below (Bottom), or you will look in along the X axis (Left and Right) or along the Y axis (Front and Back).

We will use a Southeast Isometric view. The isometric views all rotate the XY axes plus or minus 45 degrees and take you up 30 degrees out of the XY plane. It is simple if you imagine a compass. The lower right quadrant is the southeast. In a southeast isometric view, you will be looking in from this quadrant and down at a 30 degree angle. Try it.

> Select "SE Isometric" from the submenu.

The menu will disappear and the screen will be redrawn to show the view shown in *Figure 9-3*. Notice how the grid and the coordinate system icon have altered to show

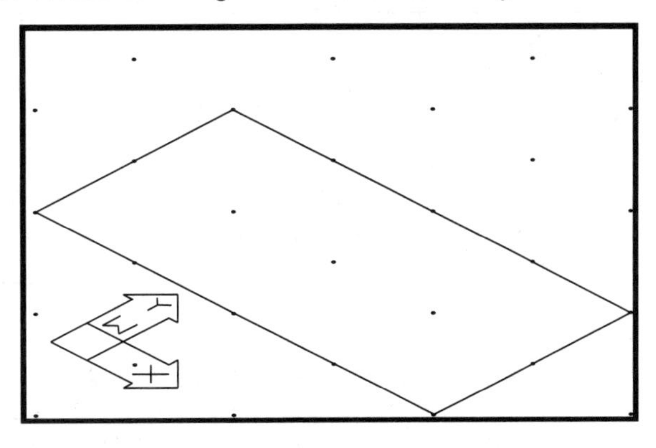

Figure 9-3

our current orientation. These visual aids are extremely helpful in viewing 3D objects on the flat screen and imagining them as if they were positioned in space.

At this point, you may wish to experiment with the other views in the 3D Viewpoint Preset submenu. You will probably find the isometric views most interesting at this point. Pay attention to the grid and the icon as you switch views. Variations of the icon you may encounter here and later on are shown in *Figure 9-4*. With some views, you will have to think carefully and watch the icon to understand which way the object is being presented.

When you are done experimenting, be sure to return to the southeast isometric view as in *Figure 9-3*. We will use this view frequently throughout this chapter.

Whenever you change viewpoints, AutoCAD displays the drawing extents so that the object fills the screen and is as large as possible. Often, you will need to zoom out a bit to get some space to work in. This is easily done using the "Scale(X)" option of the ZOOM command.

> Type "z" or select the Zoom Scale tool from the Zoom flyout on the Standard toolbar, as shown in *Figure 9-5*.

AutoCAD prompts:

All/Center/Dynamic/Extents/Left/Previous/Vmax/Window/<Scale(X/XP)>:

> In response to the ZOOM prompt, type ".5x".

Don't forget the "x". This tells AutoCAD to adjust and redraw the display so that objects appear half as large as before.

Your screen will be redrawn with the rectangle at half its previous magnification.

ICON	DESCRIPTION
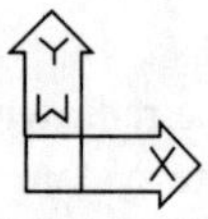	WCS (WORLD COORDINATE SYSTEM) "W" appears on "Y" arm
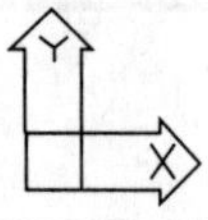	UCS (USER COORDINATE SYSTEM) NO "W" appears on "Y" arm
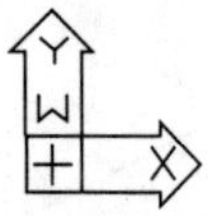	+ appears in box and "W" appears on "Y" arm if the current UCS is the same as the WCS
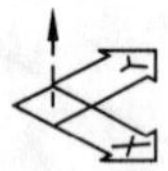 Pos. "Z"	Box appears at the base of ICON if viewing UCS from above its X–Y plane
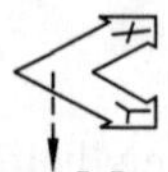 Neg. "Z"	Box is missing if viewing UCS from below its X–Y plane
	Broken pencil ICON appears if viewing direction is "EDGE ON" or near "EDGE ON" X–Y plane of current UCS

Figure 9-4

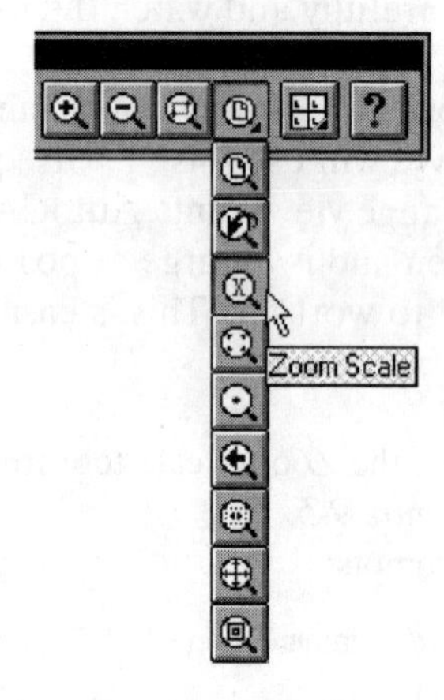

Figure 9-5

Entering 3D Coordinates

Next, we will create a copy of the rectangle placed 1.25 above it. This brings up a basic 3D problem: AutoCAD interprets all pointer device point selections as being in the XY plane, so how does one indicate a point or a displacement in the Z direction? There are three possibilities: typed 3D coordinates, X/Y/Z point filters, and object snaps. Object snap requires an object already drawn above or below the XY plane, so it will be of no use right now. We will use typed coordinates in this exercise. Later we will be using object snap as well. Point filters are not included in this book.

3D coordinates can be entered from the keyboard in the same manner as 2D coordinates. Often, this is an impractical way to enter individual points in a drawing. However, within COPY or MOVE it provides a simple method for specifying a displacement in the Z direction.

> Type "Copy" or select the Copy tool from the Modify toolbar.

AutoCAD will prompt for object selection.

> Select the complete rectangle.

> Press enter, the space bar, or the enter equivalent button on your pointing device to end object selection.

AutoCAD now prompts for the base point of a vector or a displacement value:

<Base point or displacement>/Multiple:

Typically, you would respond to this prompt and the next by showing the two end points of a vector. However, we cannot show a displacement in the Z direction by pointing. This is important for understanding AutoCAD coordinate systems. Unless an object snap is used, all points picked on the screen with the pointing device will be interpreted as being in the XY plane of the current UCS. Without an entity outside the XY plane to use in an object snap, there is no way to point to a displacement in the Z direction.

> Type "0,0,1.25".

AutoCAD now prompts:

Second point of displacement:

You can type the coordinates of another point, or press enter to tell AutoCAD to use the first entry as a displacement from (0,0,0). In this case, pressing enter will indicate a displacement of 1.25 in the Z direction and no change in X or Y.

> Press enter.

AutoCAD will create a copy of the rectangle 1.25 directly above the original. Your screen should resemble *Figure 9-6.*

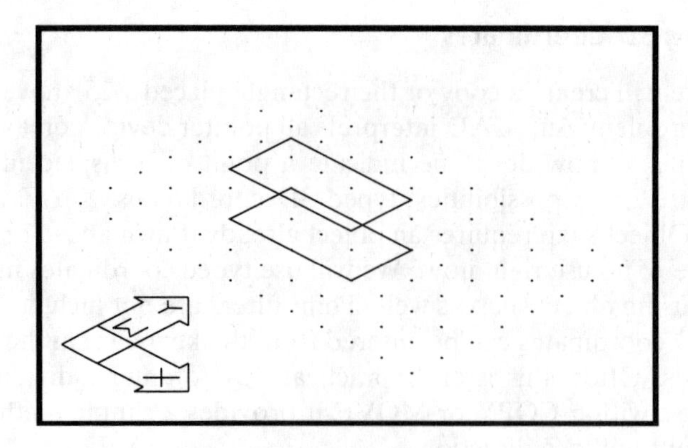

Figure 9-6

Using Object Snap

We now have two rectangles floating in space. Our next job is to connect the corners to form a wireframe box. This is done easily using "ENDpoint" object snaps. This is a good example of how object snaps allow us to construct entities not in the XY plane of the current coordinate system.

> Type "Osnap" or select "Running Object Snap . . ." from the Options pull down menu.
> Type "end" or select "Endpoint"in the dialogue box and then click on OK.

The running Endpoint object snap is now on and will affect all point selection. You will find that object snaps are very useful in 3D drawing and that the Endpoint mode can be used frequently.

Now we will draw some lines.

> Enter the LINE command and connect the upper and lower corners of the two rectangles, as shown in *Figure 9-7.* (We have removed the grid for clarity, but you will probably want to leave yours on.)

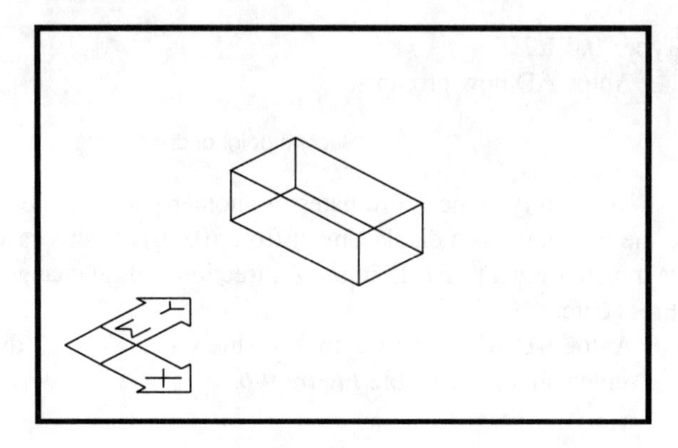

Figure 9-7

Before going on, pause a moment to be aware of what you have drawn. The box on your screen is a true wireframe model. It is a 3D model that can be turned, viewed, and plotted from any point in space. It is not, however, a solid model or a surface model. It is only a set of lines in 3D space. Removing hidden lines or shading would have no effect on this model.

In the next task, you will begin to define your own coordinate systems that will allow you to perform drawing and editing functions in any plane you choose.

TASK 2: Defining and Saving User Coordinate Systems

In this task, you will begin to develop new vocabulary and techniques for working with objects in 3D space. The primary tool will be the UCS command. You will also learn to use the UCSICON command to control the placement of the coordinate system icon.

Until now, we have had only one coordinate system to work with. All coordinates and displacements have been defined relative to a single point of origin. In Task 1 we changed our point of view, but the UCS icon changed along with it so that the orientations of the X, Y, and Z axes relative to the object were retained. With the UCS command, you can define new coordinate systems at any point and any angle in space. When you do, you can use the coordinate system icon and the grid to help you visualize the planes you are working in, and all commands and drawing aids will function relative to the new system.

The coordinate system we are currently using is unique. It is called the World Coordinate System and is the one we always begin with. The "w" at the base of the coordinate system icon indicates that we are working in the World System. A User Coordinate System is nothing more than a new point of origin and a new orientation for the X, Y, and Z axes.

We will begin by defining a User Coordinate System in the plane of the top of the box, as shown in *Figure 9-8.*

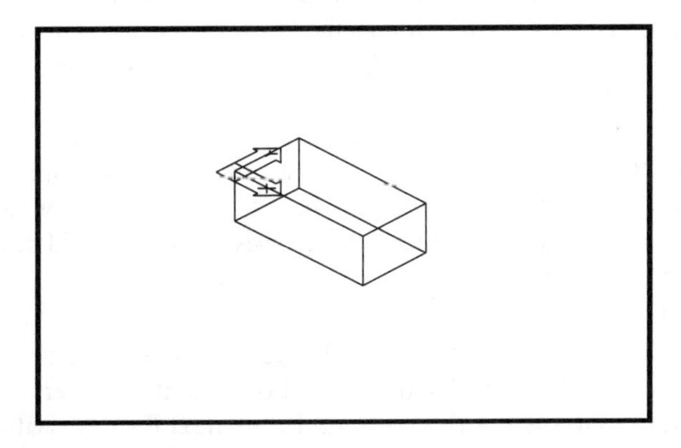

Figure 9-8

> Leave the Endpoint osnap mode on for this exercise.
> Type "UCS" or select "Set UCS" from the View pull down menu.

The UCS command gives you the following prompt:

Origin/ZAxis/3point/OBject/View/X/Y/Z/Prev/Restore/Save/Del/?/<World>:

On the pull down menu you will see the same options on a submenu. In this chapter we will explore "Origin" and "3Point". For further information, see the *AutoCAD Command Reference.*

First, we will use "Origin" to create a UCS that is parallel to the WCS.
> Type "o" or select "Origin" to specify the origin option.

AutoCAD will prompt for a new origin:

Origin point <0,0,0>:

This option does not change the orientation of the three axes. It simply shifts their intersection to a different point in space. We will use this simple procedure to define a UCS in the plane of the top of the box.
> Use the Endpoint object snap to select the top-left front corner of the box, as shown by the location of the icon in *Figure 9-8.*

You will notice that the "w" is gone from the icon. However, the icon has not moved. It is still at the lower left of the screen. It is visually helpful to place it at the origin of the new UCS, as in the figure. In order to do this, we need the UCSICON command.
> Type "Ucsicon" or select "UCS" from the Options pull down menu.

If you are typing, AutoCAD prompts:

ON/OFF/All/Noorigin/ORigin <ON>:

The first two options allow you to turn the icon on and off. The "All" option affects icons used in multiple view ports. "Noorigin" and "ORigin" allow you to specify whether you want to keep the icon in the lower left corner of the screen or place it at the origin of the current UCS. The "Follow" option on the pull down menu will cause the screen view point to follow the UCS, so that you will switch to a plan view in each new UCS as you define it. This is usually not desirable.
> Type "or" or select "Icon origin".

The icon will move to the origin of the new current UCS, as in *Figure 9-8.* With UCSICON set to origin, the icon will shift to the new origin whenever we define a new UCS. The only exception would be if the origin were not on the screen or too close to an edge for the icon to fit. In these cases, the icon would be displayed in the lower left corner again.

The "top" UCS we have just defined will make it easy to draw and edit entities that are in the plane of the top of the box and also to perform editing in planes that are parallel to it, such as the bottom. In the next task, we will begin drawing and editing using different coordinate systems and you will see how this works. For now, we will spend a little more time on the UCS command itself. We will define one more

User Coordinate System, but first we will save this one so that it can be recalled quickly later on.

> Type "UCS" or select "Set UCS" from the View menu.
This time, we will use the "Save" option.
> Type "s" or select "Save".
AutoCAD will ask you to name the current UCS so that it can be called out later:

?/Desired UCS name:

We will name our UCS "top." It will be the UCS we use to draw and edit in the top plane. This UCS would also make it easy for us to create an orthographic top view later on.
> Type "top".
The top UCS is now saved and can be recalled using the "Restore" option or by making it "current" using the UCS Control dialogue box (select "Named UCS . . ." from the View pull down menu).

Next, we will define a "front" UCS using the "3 point" option of the UCS command.

> Repeat the UCS command or select "Set UCS" from the View menu.
> Type "3" or select the "3 point" option.
AutoCAD prompts:

Origin point <0,0,0>:

In this option you will show AutoCAD a new origin point, as before, and then a new orientation for the axes as well. Notice that the default origin is the current one. If we retained this origin, we could define a UCS with the same origin and a different axis orientation.

Instead, we will define a new origin at the lower left corner of the front of the box, as shown in *Figure 9-9*.

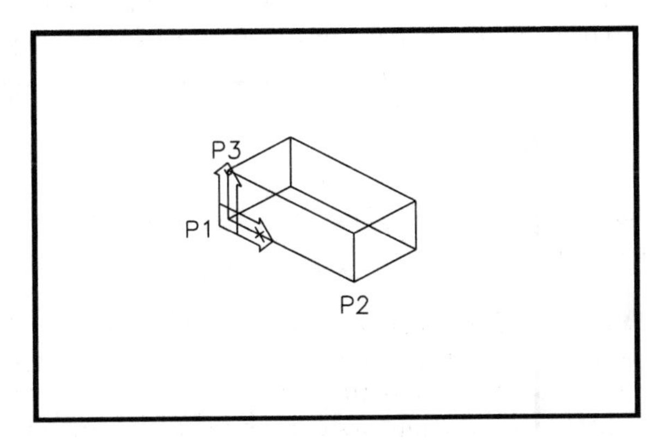

Figure 9-9

> With the Endpoint osnap on, pick point 1, as shown in the figure.

AutoCAD now prompts you to indicate the orientation of the X axis:

Point on positive portion of the X axis <1.00,0.00,-1.25>:

> Pick the right front corner of the box (point 2), as shown.

The object snap ensures that the new X axis will now align with the front of the object. AutoCAD prompts for the Y axis orientation:

Point on positive-Y portion of the UCS XY plane <0.00,1.00,-1.25>:

By definition, the Y axis will be perpendicular to the X axis, therefore, AutoCAD needs only a point that shows the plane of the Y axis and its positive direction. Because of this, any point on the positive side of the Y plane will specify the Y axis correctly. We have chosen a point that is on the Y axis itself.

> Pick point 3, as shown.

When this sequence is complete, you will notice that the coordinate system icon has rotated along with the grid and moved to the new origin as well. This UCS will be convenient for drawing and editing in the front plane of the box or editing in any plane parallel to the front, such as the back.

TASK 3: Using Draw and Edit Commands in a UCS

Now the fun begins. Using our new coordinate system, we will give the box a more interesting shape. In this task, we will cut away a slanted surface on the right side of the box. Since the planes we will be working in are parallel to the front of the box, we will be working in the "front" UCS. All our work in this task will be in this UCS.

Before going on, we need to turn off the running Endpoint osnap.

> Type "osnap" or select "Running Object Snap . . ." from the Options menu.

> Press enter at the prompt for object snap modes or select "Clear All" and "OK" from the dialogue box.

Now there will be no running object snap modes in effect.

Look at *Figure 9-10.* We will draw a line down the middle of the front (line 1) and use it to trim another line coming in at an angle from the right (line 2).

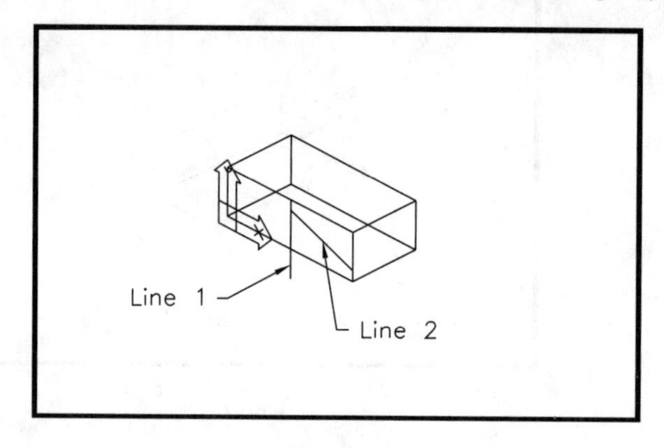

Figure 9-10

> Type "l" or select the Line tool from the Draw toolbar.
> At the "From point:" prompt, type "mid" or select the Snap to Midpoint tool from the Object Snap fl... on the Standard toolbar.
ont edge of the box.
l snap to the midpoint of the line.
tho is on (F8).
rtho works as usual, but relative to the current UCS.
nt anywhere below the box and exit the LINE command.
be trimmed later, so the exact length does not matter.

line 2 on an angle across the front. This line will become one e. Your snap setting will need to be at .25 or smaller, and or- The grid, snap, and coordinate display all work relative to the imple matter to draw in this plane.

setting and change it if necessary.
8).
command.
), pick a point .25 down from the top edge of the box on line 1, as shown in *Figure 9-10.*
> Pick a second point .25 up along the right front edge of the box.
> Exit the LINE command.
Now trim line 1.
> Type "trim" or select the Trim tool from the Modify toolbar.

Press F2 and you will see the following message in the command area:

View is not plan to UCS. Command results may not be obvious.

In the language of AutoCAD 3D, a view is plan to the current UCS if the XY plane is in the plane of the monitor display and its axes are parallel to the sides of the screen. This is the usual 2D view in which the Y axis aligns with the left side of the display and the X axis aligns with the bottom of the display. In previous chapters we always worked in plan view. In this chapter we have not been in plan view since the beginning of Task 1.

With this message, AutoCAD is warning us that boundaries, edges, and intersections may not be obvious as we look at a 3D view of an object. For example, lines that appear to cross may be in different planes. Having read the warning, we continue.

> Select line 2 as a cutting edge.
> Press enter to end cutting edge selection.
> Point to the lower end of line 1.
> Press enter to exit TRIM.

Your screen should resemble *Figure 9-11.*

Now we will copy our two lines to the back of the box. Since we will be moving out of the front plane, which is also the XY plane in the current UCS, we will require the use of Endpoint object snaps to specify the displacement vector. We will also be

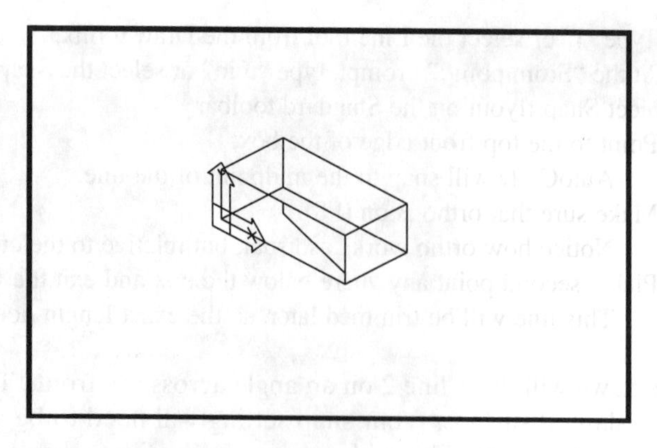

Figure 9-11

using the Endpoint osnap in the next sequence, so let's turn on the running mode again.

> Type "osnap" or select "Running Object Snap . . ." from the Options menu.
> Type "end" or select "Endpoint" and press enter or click in the "OK" box.
> Type "copy" or select the Copy tool from the Modify toolbar.
> Pick lines 1 and 2. (You may need to zoom in on the box at this point to pick line 1.)
> Press enter to end object selection.
> Use the Endpoint osnap to pick the lower front corner of the box, point 1 as shown in *Figure 9-12*.

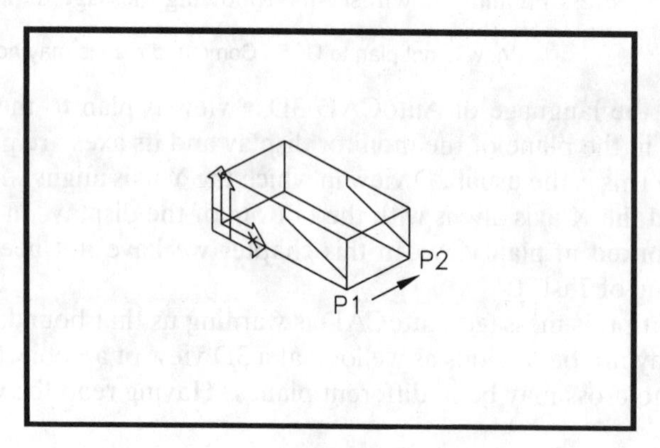

Figure 9-12

> At the prompt for a second point of displacement, use the ENDpoint osnap to pick the lower-right back corner of the box (point 2) as shown.

Your screen should now resemble *Figure 9-12*.

What remains is to connect the front and back of the surfaces we have just outlined and then trim away the top of the box. We will continue to work in the front UCS and to use Endpoint object snaps.

We will use a multiple COPY to copy one of the front-to-back edges in three places.

> Enter the COPY command.
> Pick any of the front-to-back edges for copying.
> Press enter to end object selection.
> Type "m" for "Multiple".
> Pick the front ENDpoint of the selected edge to serve as a base point of displacement.
> Pick the top end point of line 1 (point 1 in *Figure 9-13*).

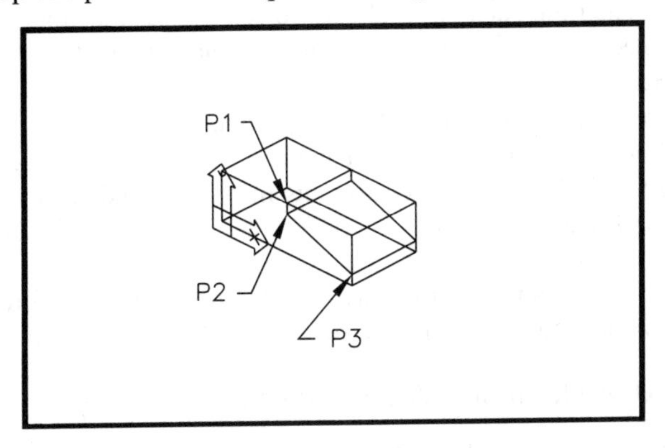

Figure 9-13

> Pick the lower end point of line 1 (point 2) as another second point.
> Pick the right end point of line 2 (point 3) as another second point.
> Press enter to exit the COPY command.
Finally, we need to do some trimming.
> Type "osnap" or select "Running Object Snap . . ." from the Options menu and turn off the running "ENDpoint" snap.
> Type "Trim" or select the Trim tool from the Modify toolbar.

For cutting edges, we want to select lines 1 and 2 and their copies in the back plane (lines 3 and 4 in *Figure 9-14*). Since this can be difficult, a quick alternative to

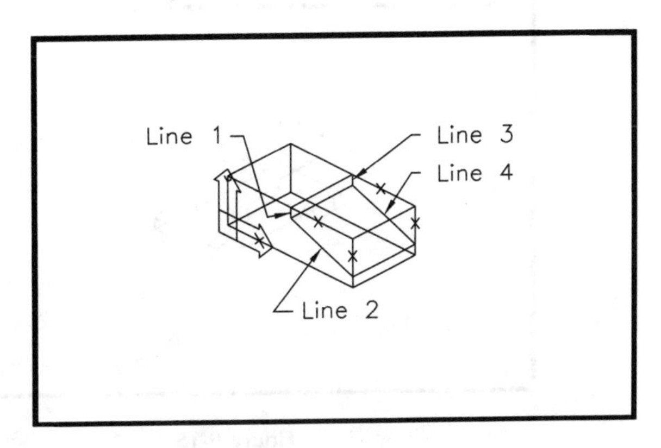

Figure 9-14

selecting these four separate lines is to use a crossing box to select the whole area. As long as your selection includes the four lines, it will be effective.
> Select the area around lines 1 and 2.

NOTE: Trimming in 3D can be tricky. Remember where you are. Edges that do not run parallel to the current UCS may not be recognized at all.

> Press enter to end cutting edge selection.
> One by one, pick the top front and top back edges to the right of the cut, and the right front and right back edges above the cut, as shown by the Xs in *Figure 9-14.* If you try to select the front to back edge, you will not be able to because it is only a point in the current UCS.
> Press enter to exit the TRIM command.
> Enter the ERASE command and erase the top edge that is left hanging in space. We must use ERASE here because this line is perpendicular to the current UCS and does not intersect any edges.

Your screen should now resemble *Figure 9-15.* This completes this wireframe figure. In the next task, we will use multiple tiled viewports and surface commands.

TASK 4: Using Multiple Tiled Viewports

A major feature needed to draw effectively in 3D is the ability to view an object from several different points of view simultaneously as you work on it. The VPORTS, or VIEWPORTS, command is easy to use and can save you from having to jump back and forth between different views of an object. Viewports can be used to place several 3D viewpoints on the screen at once. This can be a significant drawing aid. If you do not continually view an object from different points of view, it is easy to create entities that appear correct in the current view, but that are clearly incorrect from other points of view.

In this task we will divide your screen in half and define two views, so that you can visualize an object in plan view and a 3D view at the same time. As you work, re-

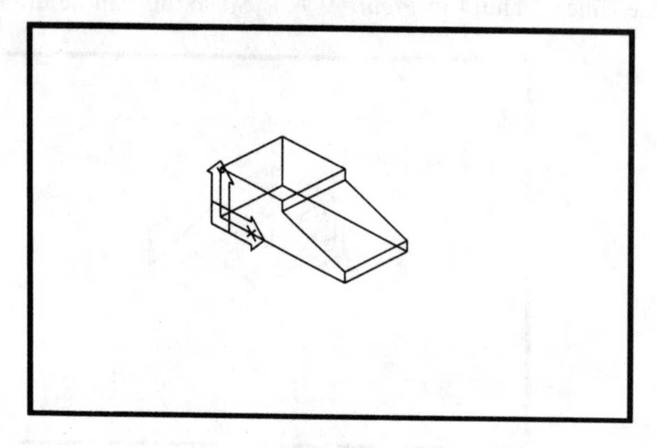

Figure 9-15

member that this is only a display command. The viewports we use in this chapter will be simple "tiled" viewports. Tiled viewports cover the complete drawing area, do not overlap, and cannot be plotted. Plotting multiple viewports can also be accomplished, but only in paper space with non-tiled viewports.

> Before going on, we will return to the World Coordinate System.

It is always advisable to keep track of where you are in relation to the WCS and to begin there.

> Type "ucs" or select "Set UCS" from the View pull down menu.

> Type "w" or select "World" from the submenu.

Now we will move to two viewports.

> Type "Vports" or select "Tiled Viewports" from the View menu.

If you are typing the command, AutoCAD will show you the following prompt; if you are on the pull down menu, you will see the same options listed on a submenu:

Save/Restore/Delete/Join/SIngle/?/2/<3>/4:

The first three options allow you to save, restore, and delete viewport configurations. "Join" allows you to reduce the number of windows so that you move from, say, four windows to three. "SIngle" is the option that returns you to a single window. "?" will get you a list of previously saved viewport configurations. The numbers 2, 3, and 4 will establish the number of different viewports you want to put on the screen.

From the pull down menu, the "Layout..." option will call the Tiled Viewport Layout dialogue box shown in *Figure 9-16*. Notice that the list on the left simply names the twelve layout options on the right. Any of these could also be specified by typing and following the sequence of command prompts.

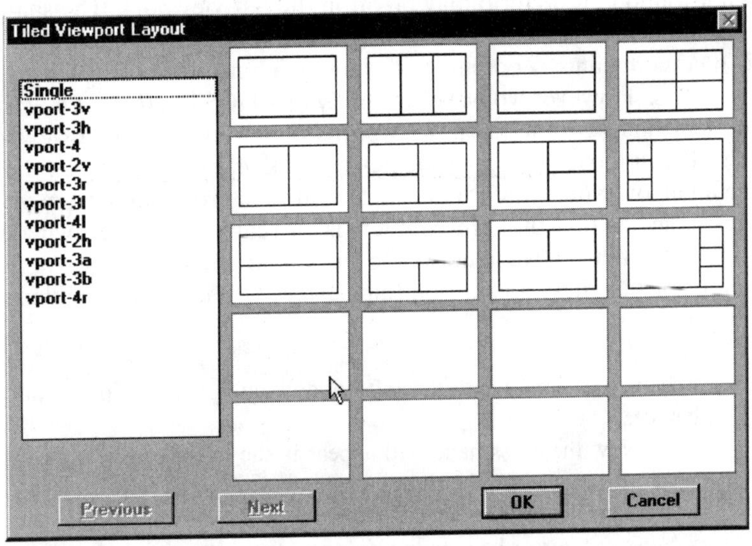

Figure 9-16

From the command line, the default is three windows with a full pane on the right and a horizontal split on the left. This is called "vport-3r" in the dialogue box.

For our purposes we will create a simple two-way vertical split, the "vport-2v" option in the dialogue box.

> Type "2" or select the box on the left end of the second row on the icon menu.

If you type the number, AutoCAD will prompt for the direction of the split:

Horizontal/<Vertical>:

> Press enter to accept the default vertical split, or click on "OK" to complete the dialogue.

Your screen will be regenerated to resemble *Figure 9-17*.

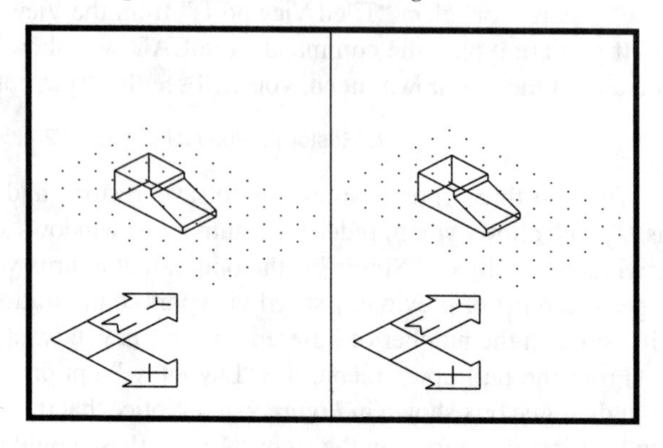

Figure 9-17

You will notice that the grid is rather small and confined to the lower part of the window and that the grid may be off in the left viewport. The shape of the viewports necessitates the reduction in drawing area. You can enlarge details as usual using the ZOOM command. Zooming and panning in one viewport will have no effect on other viewports. However, you can have only one UCS in effect at any time, so a change in the coordinate system in one viewport will be reflected in all viewports.

If you move your pointing device back and forth between the windows, you will see an arrow when you are on the left and the cross hairs when you are on the right. This indicates that the right window is currently active. Drawing and editing can be done only in the active window. To work in another window, you need to make it current by picking it with your pointing device. Often, this can be done while a command is in progress.

> Move the cursor into the left window and press the pick button on your pointing device.

Now the cross hairs will appear in the left viewport, and you will see the arrow when you move into the right viewport.

> If the grid is off in your left viewport, you may want to turn it on at this point.

> Move the cursor back to the right and press the pick button again.

This will make the right window active again.

There is no value in having two viewports if each is showing the same thing, so our next job will be to change the viewpoint in one of the windows. We will leave the window on the right with a southeast plan view and switch the left window to a plan view.

> Click in the left viewport to make it active.
> Type "plan" or select "3D Viewpoint Presets," then "Plan View" from the pull down.
> Type "w" or select "World" from the pull down submenu. In this case, "c" or "Current" will also work.

Your screen should now be redrawn with a plan view in the left viewport, as shown in *Figure 9-18.*

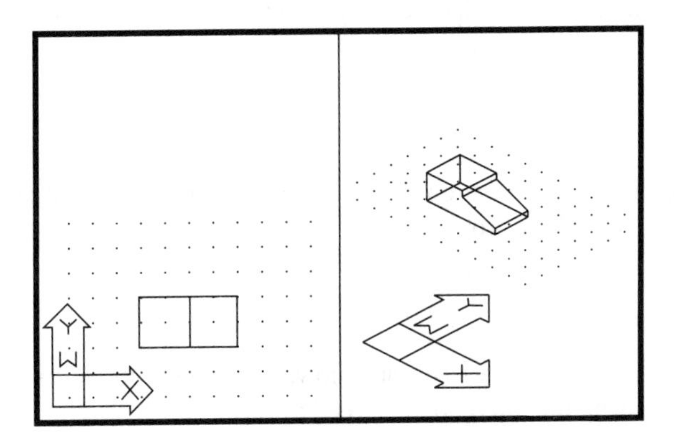

Figure 9-18

Once you have defined viewports, any drawing or editing done in the active viewport will appear in all the viewports. As you draw, watch what happens in both viewports.

TASK 5: Creating Surfaces with 3DFACE

3DFACE creates triangular and quadrilateral surfaces. 3D faces are built by entering points in groups of three or four to define the outlines of triangles or quadrilaterals. The surface of a 3D face is not shown on the screen, but it is recognized by the HIDE, SHADE, and RENDER commands.

In this exercise, we will add a surface to the top of the wedge block. You may eventually want a number of layers specifically defined for faces and surfaces, but this will not be necessary for the current exercise.

> Type "3dface" or select the 3D Face tool from the Surfaces toolbar, as shown in *Figure 9-19.*

AutoCAD will prompt:

First point:

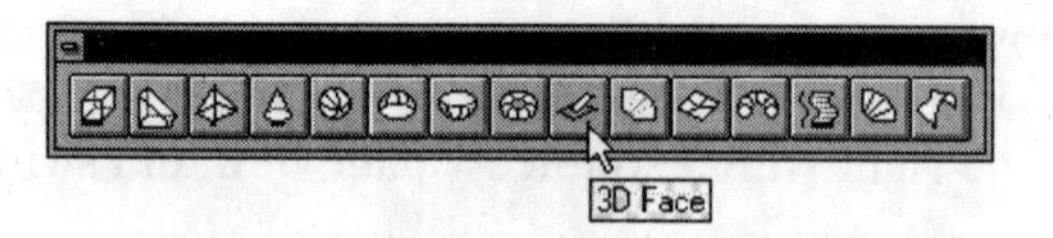

Figure 9-19

You can define points in either of the two viewports. In fact, you can even switch viewports in the middle of the command.

> Using an Endpoint object snap, pick a point similar to point 1, as shown in *Figure 9-20*.

AutoCAD prompts:

Second point:

> Using another Endpoint object snap, pick a second point, moving around the perimeter of the face, as shown.

AutoCAD prompts:

Third point:

> Using an Endpoint object snap, pick a third point as shown.

AutoCAD prompts:

Fourth point:

NOTE: If you pressed enter now, AutoCAD would draw the outline of a triangular face, using the three points already given.

> Using an Endpoint object snap, pick the fourth point of the face.

AutoCAD draws the fourth edge of the face automatically when four points have been given, so it is not necessary to complete or close the rectangle.

AutoCAD will continue to prompt for third and fourth points so that you can draw a series of surfaces to cover an area with more than four edges. Keep in mind,

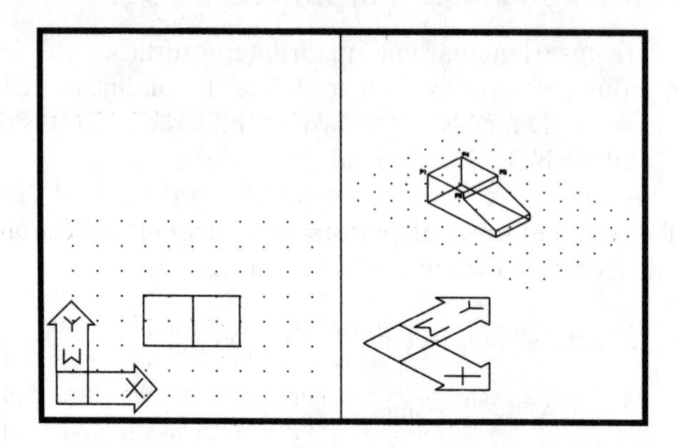

Figure 9-20

however, that drawing faces in series is only a convenience. The result is a collection of independent three- and four-sided faces.

> Press enter to exit the 3DFACE command.

In the next task, we will demonstrate the HIDE command.

TASK 6: Removing Hidden Lines with HIDE

The HIDE command is easy to execute. However, execution may be slow in large drawings, and careful work may be required to create a drawing that hides the way you want it to. This is a primary objective of surface modeling. When you've got everything right, HIDE will temporarily remove all lines and objects that would be obstructed in the current view, resulting in a more realistic representation of the object in space. A correctly surfaced model can also be used to create a shaded or rendered image. Hiding has no effect on wireframe drawings, since there are no surfaces to obstruct lines behind them.

> Make the right viewport active.
> Type "hide" (there is a Hide tool on the Render toolbar, but it is not worth opening the toolbar for this one command).

Your screen should be regenerated to resemble *Figure 9-21.* Notice that the object has a "top" on it, the surface you created in the last task.

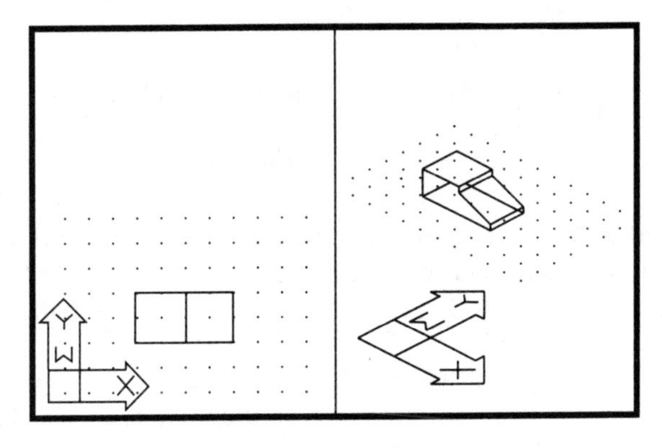

Figure 9-21

Following are some important points about hidden line removal which you should read before continuing:

1. Hidden line removal can be done from the plot configuration dialogue box. However, due to the time involved and the difficulty of getting a hidden view just right, it is usually better to experiment on the screen first, then plot with hidden lines removed when you know you will get the image you want.
2. Whenever the screen is regenerated, hidden lines are returned to the screen.

3. Layer control is important in hidden line removal. Layers that are frozen are ignored by the HIDE command, but layers that are off are treated like layers that are visible. This can, for example, create peculiar blank spaces in your display if you have left surfaces or solids on a layer that is off at the time of hidden line removal.

TASK 7: Using 3D Polygon Mesh Commands

3DFACE can be used to create simple surfaces. However, most surface models require large numbers of faces to approximate the surfaces of real objects. Consider the number of faces in *Figure 9-22*. Obviously you would not want to draw such an image one face at a time.

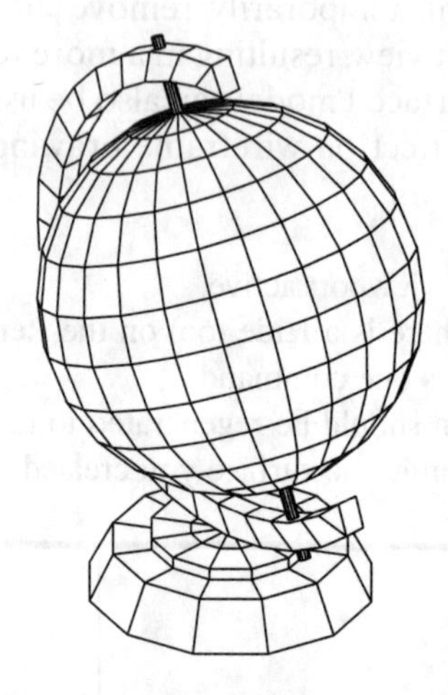

Figure 9-22

AutoCAD includes a number of commands that make the creation of some types of surfaces very easy. These powerful commands create 3D polygon meshes. Polygon meshes are made up of 3D faces and are defined by a matrix of vertices.

> To begin this task, ERASE the wedge block from the previous tasks and draw an arc and a line below it as shown in *Figure 9-23*. Exact sizes and locations are not important. The entities may be drawn in either viewport.

Now we will define some 3D surfaces using the arc and line you have just drawn.

TABSURF

The first surface we will draw is called a "tabulated surface." In order to use the TABSURF command, you need a line or curve to define the shape of the surface and a vector to show its size and direction. The result is a surface generated by repeating the shape of the original curve at every point along the path specified by the vector.

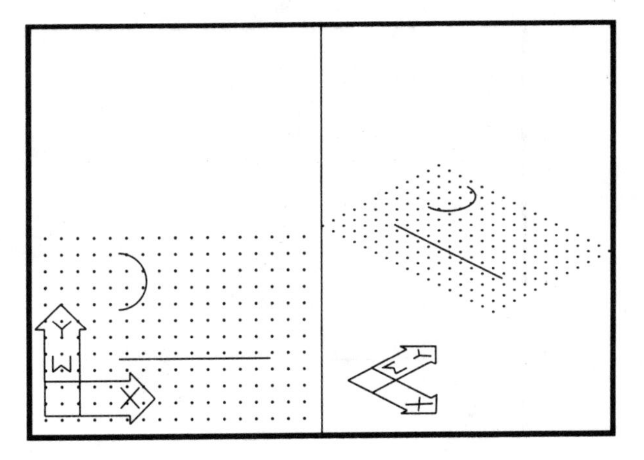

Figure 9-23

> Type "Tabsurf" or select the Extruded Surface tool from the Surfaces toolbar, as shown in *Figure 9-24.*

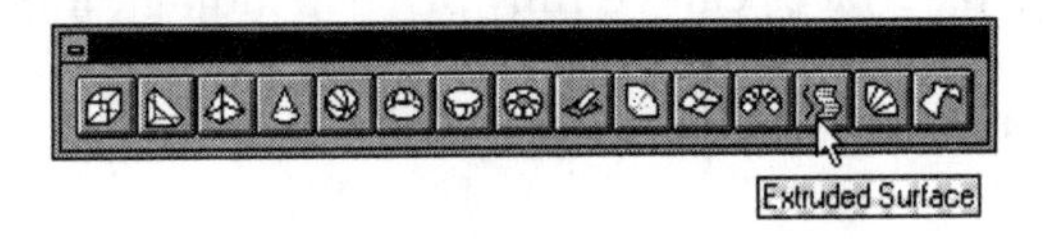

Figure 9-24

AutoCAD will prompt:

Select path curve:

The path curve is the line or curve that will determine the shape of the surface. In our case, it will be the arc.

> Pick the arc.

AutoCAD will prompt for a vector:

Select direction vector:

We will use the line. Notice that the vector does not need to be connected to the path curve. Its location is not significant, only its direction and length.

There is an oddity here to watch out for as you pick the vector. If you pick a point near the left end of the line, AutoCAD will interpret the vector as extending from left to right. Accordingly, the surface will be drawn to the right. By the same token, if your point is near the right end of the line, the surface will be drawn to the left. Most of the time, you will avoid confusion by picking a point on the side of the vector nearest the curve itself.

> Pick a point on the left side of the line.

Your screen will be redrawn to resemble *Figure 9-25.*

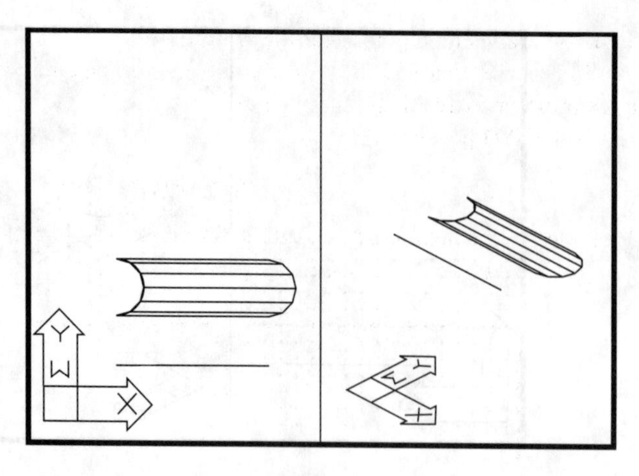

Figure 9-25

Notice that this is a flat surface even though it looks 3D. This may not be clear in the plan view, but is more apparent in the 3D view.

Tabulated surfaces can be fully 3D, depending on the path and vector chosen to define them. In this case we have an arc and a vector that are both entirely in the XY plane, so the resulting surface is also in that plane.

RULESURF

TABSURF is useful in defining surfaces that are the same on both ends, assuming you have one end and a vector. Often, however, you have no vector, or you need to draw a surface between two different paths. In these cases, you will need the RULESURF command.

For example, what if we need to define a surface between the line and the arc? Let's try it.

> Type "u" to undo the last tabulated surface, then execute a REDRAW or REDRAWALL.

It is not absolutely necessary to do the REDRAW, but it makes it easier to pick the arc.

> Type "Rulesurf" or select the Ruled Surface tool from the Surfaces toolbar, as shown in *Figure 9-26*.

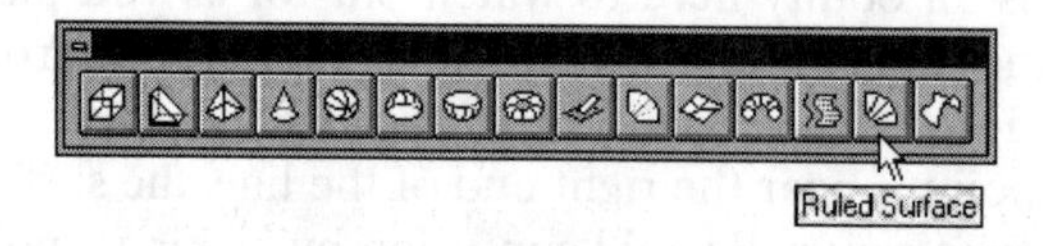

Figure 9-26

The first prompt is:

Select first defining curve:

> Pick the arc, using a point near the bottom.

Remember that you must pick points on corresponding sides of the two defining curves in order to avoid an hourglass effect.

AutoCAD prompts:

```
Select second defining curve:
```

> Pick the line, using a point near the left end.

Your screen should resemble *Figure 9-27.*

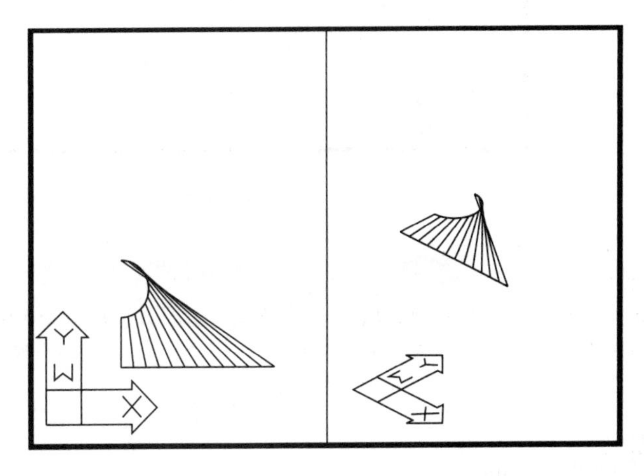

Figure 9-27

Again, notice that this surface is within the XY plane. Ruled surfaces may be drawn just as easily between curves that are not coplanar.

EDGESURF

TABSURF creates surfaces that are the same at both ends and move along a straight line vector. RULESURF draws surfaces between any two boundaries. There is a third command, EDGESURF, which draws surfaces that are bounded by four curves. Edge-defined surfaces have a lot of geometric flexibility. The only restriction is that they must be bounded on all four sides. That is, they must have four edges that touch.
In order to create an EDGESURF, we need to undo our last ruled surface and add two more edges.

> Type "u" and execute a REDRAW or REDRAWALL.
> Add a line and an arc to your screen, as shown in *Figure 9-28.*

Remember, you can draw in either viewport.

> Type "Edgesurf" or select the Edge Surface tool from the Surfaces toolbar, as shown in *Figure 9-29* (note that this is not the same as the Edge tool on the same toolbar).

AutoCAD will prompt for the four edges of the surface, one at a time:

```
Select edge 1:
```

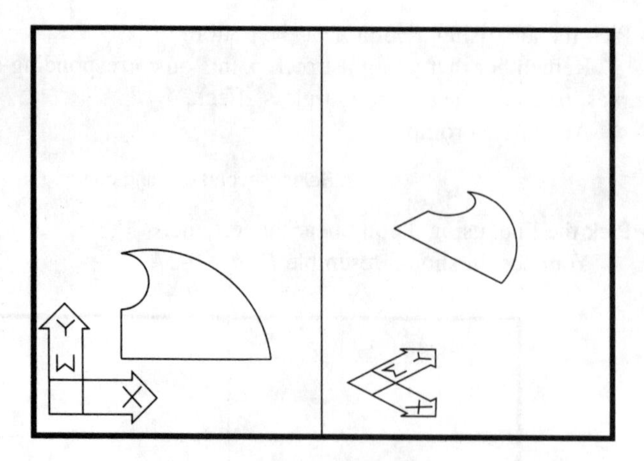

Figure 9-28

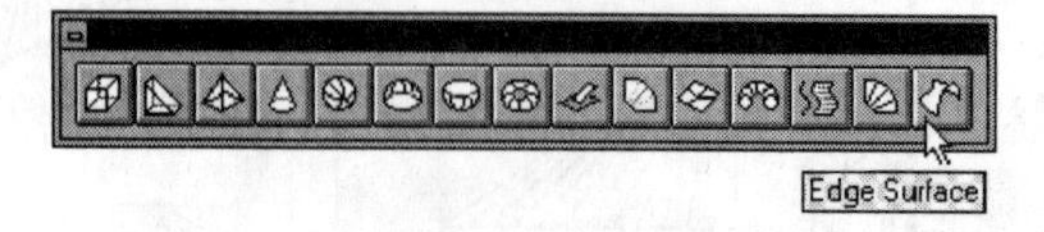

Figure 9-29

> Pick the smaller arc.

AutoCAD prompts:

Select edge 2:

> Pick the larger arc.

AutoCAD prompts:

Select edge 3:

> Pick the longer line.

AutoCAD prompts:

Select edge 4:

> Pick the smaller line.

Your screen should now resemble *Figure 9-30.*

REVSURF

We have one more 3D polygon mesh command to explore, and this one is probably the most impressive of all. REVSURF creates surfaces by spinning a curve through a given angle around an axis of revolution. Just as tabulated surfaces are spread along a linear path, surfaces of revolution follow a circular or arc-shaped path. As a result, surfaces of revolution are always fully three-dimensional, even if their defining geometry is in a single plane, as it will be here.

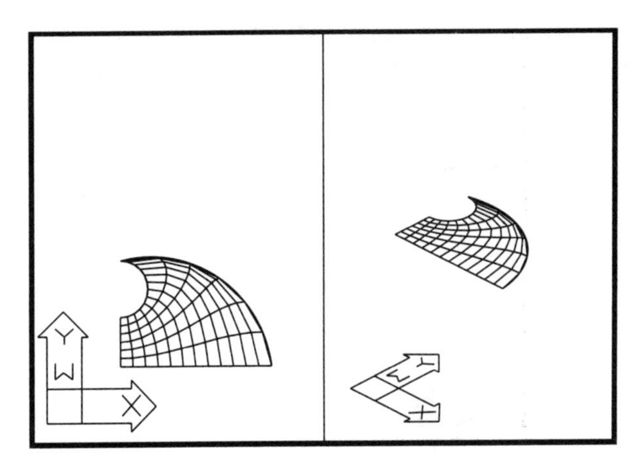

Figure 9-30

> In preparation for this exercise, undo the EDGESURF and REDRAW (or REDRAWALL) your screen so that it resembles *Figure 9-28* again.
We will use REVSURF to create a surface of revolution.
> Type "revsurf" or select the Revolved Surface tool from the Surfaces toolbar, as shown in *Figure 9-31*.

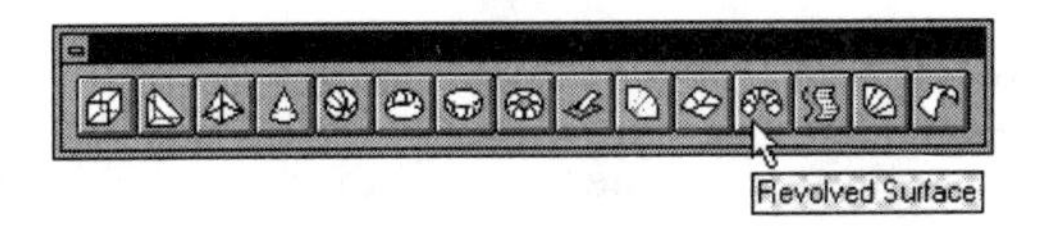

Figure 9-31

AutoCAD needs a path curve and an axis of revolution to define the surface. The first prompt is:

Select path curve:

> Pick the smaller arc.
AutoCAD prompts:

Select axis of revolution:

> Pick the smaller line.
AutoCAD now needs to know whether you want the surface to begin at the curve itself or somewhere else around the circle of revolution:

Start angle <0>:

The default is to start at the curve.
> Press enter to begin the surface at the curve itself.
AutoCAD prompts:

Included angle (+=ccw, -=cw) <Full circle>:

Entering a positive or negative degree measure will cause the surface to be drawn around an arc rather than a full circle. The default will give us a complete circle.
> Press enter.

Your screen should be drawn to resemble *Figure 9-32*.

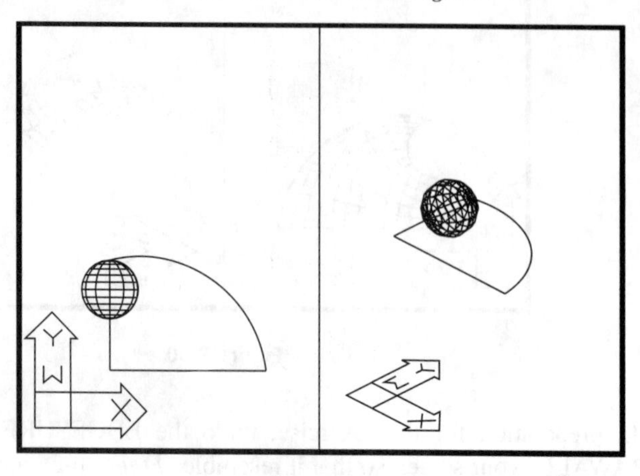

Figure 9-32

Next we will move onto solid modeling, probably the most powerful type of 3D drawing.

TASK 8: Creating Solid BOXes and WEDGEs

Solid modeling requires a somewhat different type of thinking than any of the work you have done so far. Instead of focusing on lines and arcs, edges and surfaces, you will need to imagine how 3D objects might be pieced together by combining or subtracting basic solid shapes, called "primitives." This building block process includes joining, subtracting, and intersecting operations. A simple washer, for example, could be made by cutting a small cylinder out of the middle of a larger cylinder. In AutoCAD solid modeling you can begin with a flat outer cylinder, then draw an inner cylinder with a smaller radius centered at the same point, and then subtract the inner cylinder from the outer, as illustrated in *Figure 9-33*.

This operation, which uses the SUBTRACT command, is the equivalent of cutting a hole and is one of three Boolean operations (after the mathematician George Booles) used to create composite solids. UNION joins two solids to make a new solid, and INTERSECT creates a composite solid in the space where two solids overlap.

In this exercise, you will create a composite solid from the union and subtraction of several solid primitives. Primitives are 3D solid building blocks—boxes, cones, cylinders, spheres, wedges, and torus. They all are regularly shaped and can be defined by specifying a few points and distances.

For this exercise, we will return to a single 3D viewpoint.

> Erase objects left from the last task.
> Make the right viewport active.

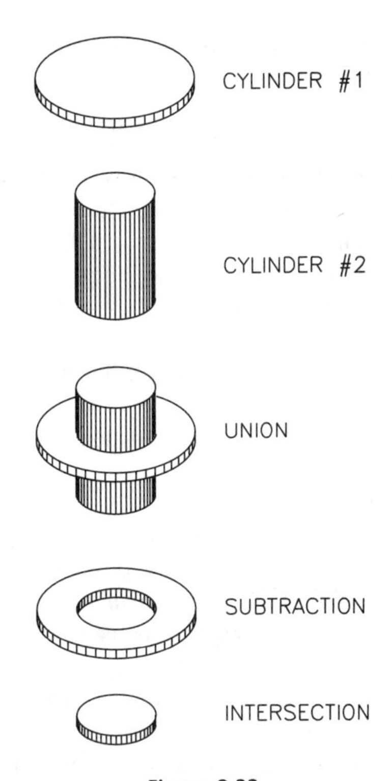

Figure 9-33

> Select Tiled Viewports from the View pull down menu.
> From the submenu, select "1 Viewport".

We are now ready to create new solid objects.

> Type "box" or select the Box Corner tool from the Box flyout on the Solid toolbar, as illustrated in *Figure 9-34*.

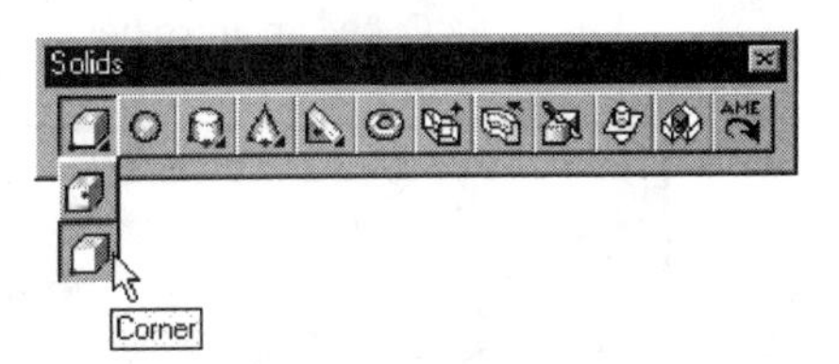

Figure 9-34

In the command area you will see the following prompt:

Center/<Corner of box> <0,0,0>:

"Center" allows you to begin defining a box by specifying its center point. Here we will use the "Corner of box" option to begin drawing a box in the baseplane of the current UCS. We will draw a box with a length of 4, width of 3, and height of 1.5.

> Pick a corner point similar to point 1 in *Figure 9-35.*
AutoCAD will prompt:

Cube/Length/<Other corner>:

With the "Cube" option you can draw a box with equal length, width, and height simply by specifying one distance. The "Length" option will allow you to specify length, width, and height separately. If you have simple measurements that fall on snap points, as we do, you can show the length and width at the same time by picking the other corner of the base of the box (the default method).

> Move the cross hairs over 4 in the X direction and up 3 in the Y direction to point 2, as shown in the figure.
Notice that "length" is measured along the X axis, and "width" is measured along the Y axis.
> Pick point 2 as shown.
Now AutoCAD prompts for a height. "Height" is measured along the Z axis. As usual, you cannot pick points in the Z direction unless you have objects to snap to. Instead, you can type a value or show a value by picking two points in the XY plane.
> Type "1.5" or pick two points 1.5 units apart.
Your screen should resemble *Figure 9-35.*
Next we will create a solid wedge. The process will be exactly the same, but there will be no "Cube" option.
> Type "Wedge" or select the Wedge Corner tool from the Wedge flyout on the Solids toolbar.
AutoCAD prompts:

Center/<Corner of wedge> <0,0,0>:

> Pick to front corner of the box, as shown in *Figure 9-36.*
As in the BOX command, AutoCAD prompts for a cube, length, or the other corner:

Cube/Length/<other corner>:

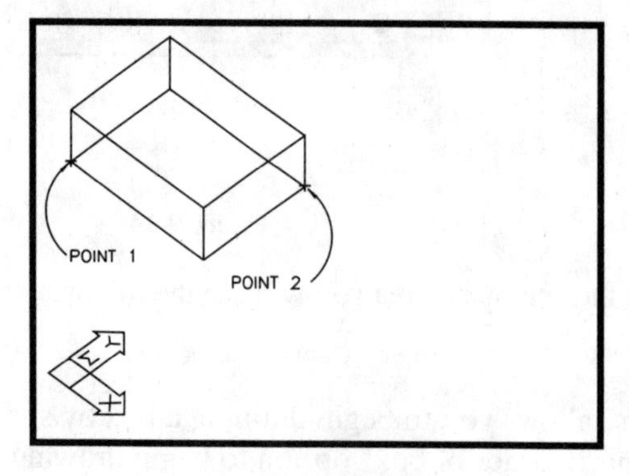

Figure 9-35

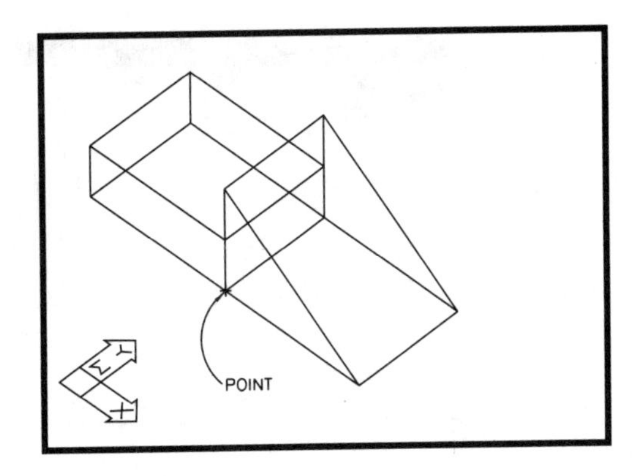

Figure 9-36

This time, let's use the length and width option.

> Type "L".

The rubber band will disappear and AutoCAD will prompt for a length, which you can define by typing a number or showing two points.

> Type "4" or show a length of 4.00 units.

AutoCAD now prompts for a width. Remember: length is measured in the X direction, and width is measured in the Y direction.

> Type "3" or show a width of 3 units.

AutoCAD prompts for a height.

> Type "3" or show a distance of 3 units.

AutoCAD will draw the wedge you have specified. Notice that a wedge is simply half a box, cut along a diagonal plane.

Your screen should resemble *Figure 9-36*. Although the box and the wedge appear as wireframe objects, they are really quite different, as you will find. In Task 9 we will join the box and the wedge to form a new composite solid.

TASK 9: Creating the UNION of Two Solids

Unions are simple to create and usually easy to visualize. The union of two objects is an object that includes all points that are on either of the objects. Unions can be performed just as easily on more than two objects. The union of objects can be created even if the objects have no points in common (i.e., they do not touch or overlap).

Although the box and wedge are adjacent, there are still two distinct solids on the screen; with UNION we can join them.

> Type "Union" or select the Union tool from the Explode flyout on the Modify toolbar, as shown in *Figure 9-37*.

Now AutoCAD will prompt you to select objects.

> Point or use a crossing box to select both objects.

> Press enter to end object selection.

Your screen should resemble *Figure 9-38*.

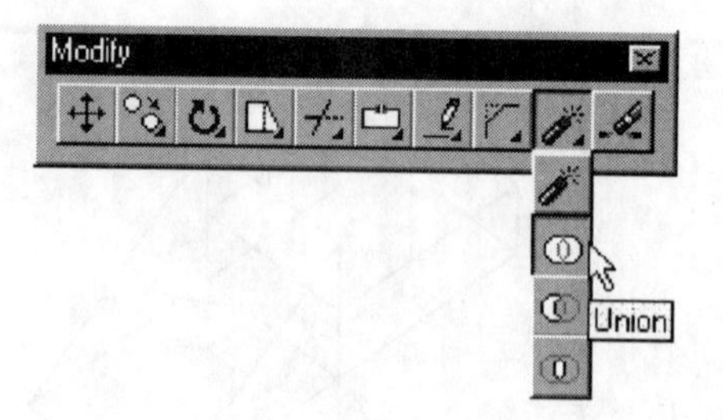

Figure 9-37

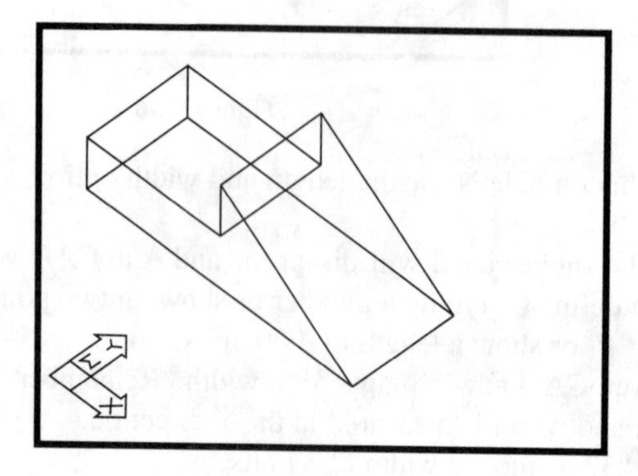

Figure 9-38

TASK 10: Working Above the XY Plane Using Elevation

In this task, we will draw two more solid primitives while demonstrating the use of elevation to position objects above the XY plane of the current UCS. Changing elevations simply adds a single Z value to all new objects as they are drawn and can be used as an alternative to creating a new UCS. With an elevation of 1.00, for example, new objects would be drawn 1.00 above the XY plane of the current UCS. You can also use a thickness setting to create 3D objects, but these will be created as mesh objects, not solids.

We will begin by drawing a third box positioned on top of the first box. Later we will move it, copy it, and subtract it to form a slot in the composite object.

> Type "Elev".

AutoCAD prompts:

New current elevation <0.00>:

The elevation is always set at 0 unless you specify otherwise.

> Type "1.5"

This will bring the elevation up 1.5 out of the XY plane, putting it even with the top of the first box you drew.

AutoCAD now prompts for a new current thickness. Thickness does not apply to solid objects because they have their own thickness.

> Press enter to retain 0.00 thickness.

This brings us back to the command prompt. If you watch closely, you will see that the grid has also moved up into the new plane of elevation.

> Type "Box" or select the Box Corner tool from the Solids toolbar.

> Pick the upper left corner of the box, point 1 in *Figure 9-39.*

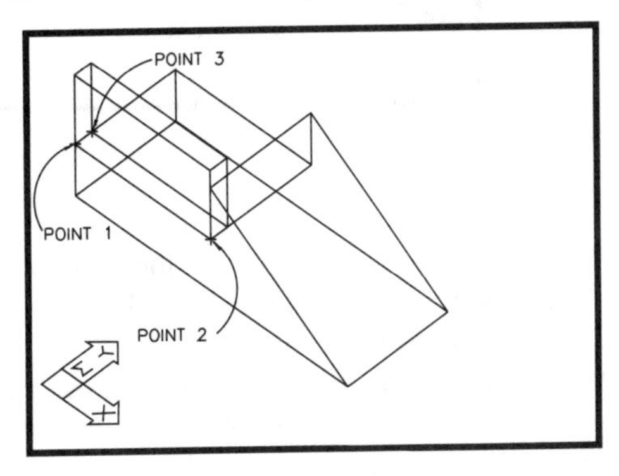

Figure 9-39

> Type "L" to initiate the "length" option.

> Type "4" or pick two points (points 1 and 2 in the figure) to show a length of 4 units.

> Type ".5" or pick two points (points 1 and 3) to show a width of .50 units.

> Type "2" or pick two points to show a height of 2 units.

Your screen should resemble *Figure 9-39.* Notice how you were able to pick points on top of the box because of the change in elevation just as if we had changed coordinate systems. Before going on, you should return to 0.00 elevation.

> Type "elev".

> Type "0".

> Press enter to retain 0.00 thickness.

TASK 11: Creating Composite Solids with SUBTRACT

SUBTRACT is the logical opposite of UNION. In a union operation, all the points contained in one solid are added to the points contained in other solids to form a new composite solid. In a subtraction, all points in the solids to be subtracted are removed from the source solid. A new composite solid is defined by what is left.

In this exercise, we will use the objects already on your screen to create a slotted wedge. First we need to move the thin upper box into place, then we will copy it to create a longer slot, and finally we will subtract it and its copy from the union of the box and wedge.

Before subtracting, we will move the box to the position shown in *Figure 9-40.*

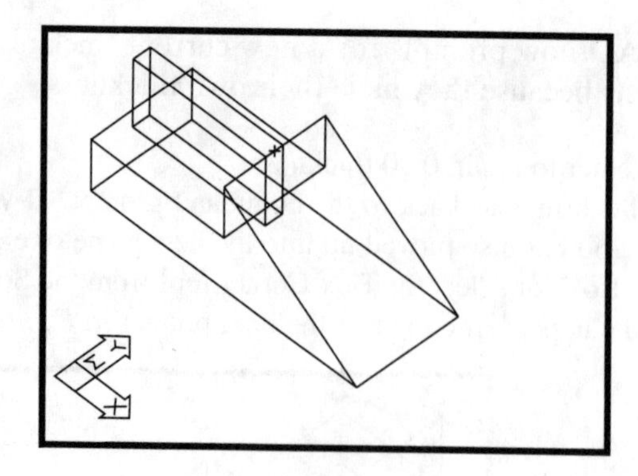

Figure 9-40

> Type "m" or select the Move tool from the Modify toolbar.
> Select the narrow box drawn in the last task.
> Press enter to end object selection.
> At the "Base point or displacement" prompt, use a midpoint object snap to pick the midpoint of the top right edge of the narrow box. (If you have trouble making this work, zoom in on the object.)
> At the "Second point of displacement" prompt, use another midpoint object snap to pick the top edge of the wedge, as shown in *Figure 9-40.*

This will move the narrow box over and down. If you were to perform the subtraction now, you would create a slot but it would only run through the box, not the wedge. We can create a longer slot by copying the narrow box over to the right using grips.

> Select the narrow box.
> Pick any of the eight grips.
> If ortho is off, turn it on (F8).

Now, if you move the cursor in the x direction you will see a copy of the box moving with you.

> Type "c" or select "Copy".
> Move the cursor between 2.00 and 4.00 units to the right and press the pick button.

If you don't go far enough, the slot will be too short. If you go past 4.00, the slot will be interrupted by the space between the box and the copy.

> Press enter to leave the grip edit mode.

Your screen will resemble *Figure 9-41.*

> Type "Subtract" or select the Subtract tool from the Explode flyout on the Modify toolbar, as shown in *Figure 9-42.*

AutoCAD asks you to select objects to subtract from first (you may have to press F2 to see this in the text window):

Select solids and regions to subtract from . . .
Select objects:

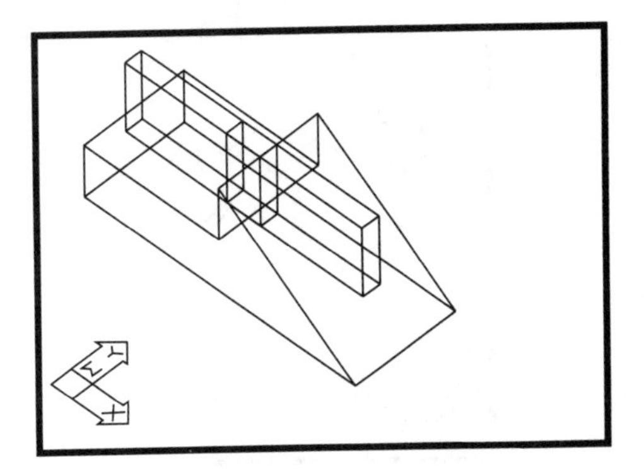

Figure 9-41

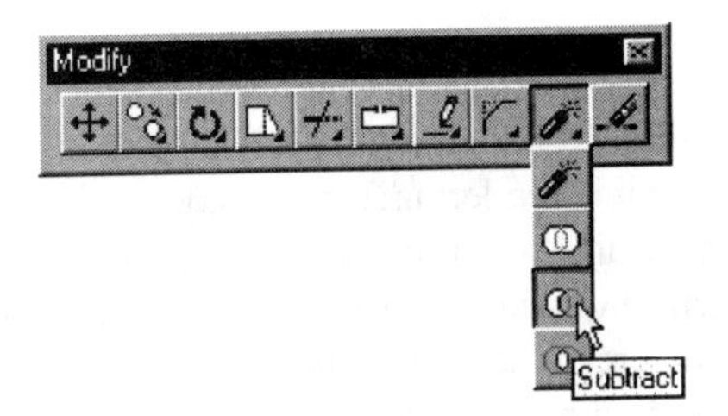

Figure 9-42

> Pick the composite of the box and the wedge.
> Press enter to end the selection of source objects.

AutoCAD prompts for objects to be subtracted:

Select solids and regions to subtract...
Select objects:

> Pick the two narrow boxes.
> Press enter to end selection.

Your screen will resemble *Figure 9-43*.

TASK 12: Shading and Rendering

There are many ways to present a solid object. Once drawn you can view it from any angle, create a perspective view using the DVIEW command, remove hidden lines, or create dramatic shaded or rendered images. Before leaving, try these three simple procedures. We hope that these will whet your appetite and that you will go on to learn much more about 3D imaging in AutoCAD.

> To remove hidden lines, type "hide".

This will produce a simple hidden line image that clearly shows how AutoCAD is recognizing the surfaces of this solid object.

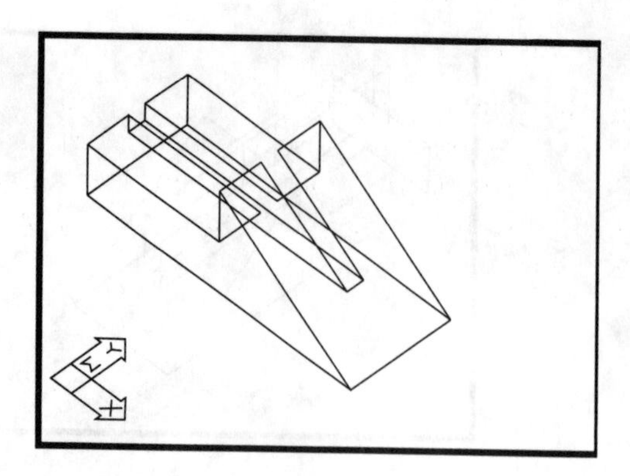

Figure 9-43

> To create a simple shaded image, type "shade".
This will produce a shaded image similar to *Figure 9-44.*
> To create a simple rendering, type "render". This will call up the Render dialogue box.
> In the Render dialogue box, select "Render Scene" in the lower left corner.

This will produce a simple rendered image similar to *Figure 9-45.* In a rendered image you can create and control effects to simulate light, shadow, and different materials. Although lighting is beyond the scope of this introduction to AutoCAD, we would encourage you to explore the LIGHT command, which allows you to add three different types of lighting to rendered images, and the RMAT and MATLIB commands which allow you to create the effect of different materials. Good luck!

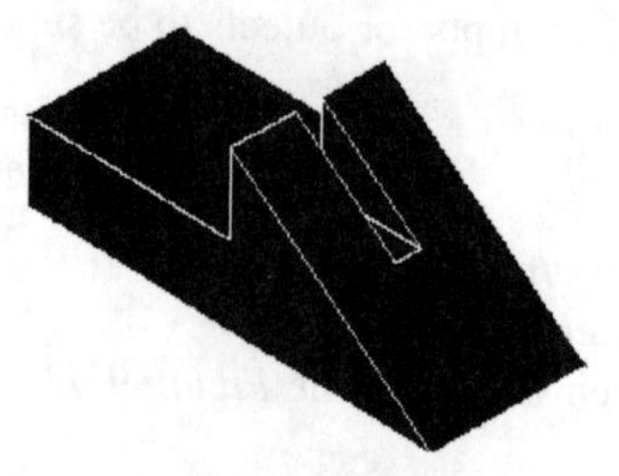

Figure 9-44

Figure 9-45

TASKS 13, 14, and 15

The drawings that follow will give you the opportunity to explore each of the 3D model types we have explored in this chapter. Drawing 9-1 is a wireframe model. Drawing 9-2 is an experiment in creating revolved surfaces. Drawing 9-3 is a solid model. In all of these drawings, you should work at gaining facility with User Coordinate Systems and 3D space.

DRAWING 9-1: CLAMP

This drawing is similar to the one you did in the chapter. Two major differences are that it is drawn from a different viewpoint and that it includes dimensions in the 3D view. This will give you additional practice in defining and using User Coordinate Systems. Your drawing should include dimensions, border, and title.

DRAWING SUGGESTIONS

> We drew the outline of the clamp in a horizontal position and then worked from a southeast isometric or front, left, top point of view.

> Begin in WCS plan view drawing the horseshoe-shaped outline of the clamp. This will include fillets on the inside and outside of the clamp. The more you can do in plan view before copying to the top plane, the less duplicate editing you will need to do later.

> When the outline is drawn, switch to a southeast isometric view.

> COPY the clamp outline up 1.50.

> Define User Coordinate Systems as needed, and save them whenever you are ready to switch to another UCS. You will need to use them in your dimensioning.

> The angled face and the slot can be drawn just as in the chapter. The Filleted surfaces are drawn using the RULESURF command.

Dimensioning in 3D

The trick to dimensioning a 3D object is that you will need to restore the appropriate UCS for each set of dimensions. Think about how you want the text to appear. If text is to be aligned with the top of the clamp (i.e., the 5.75 overall length), you will need to draw that dimension in a "top" UCS; if it is to align with the front of the object (the 17 degree angle and the 1.50 height), draw it in a "front" UCS, and so forth.

> Define a UCS with the "View" option in order to add the border and title. Type "UCS", then "v". This creates a UCS aligned with your current viewing angle.

Setting Surftab1

Notice that there are 16 lines defining the RULESURF fillets in this drawing, compared to six in the chapter. This is controlled by the setting of the variable Surftab1. You can change it by typing "Surftab1" and entering "16" for the new value.

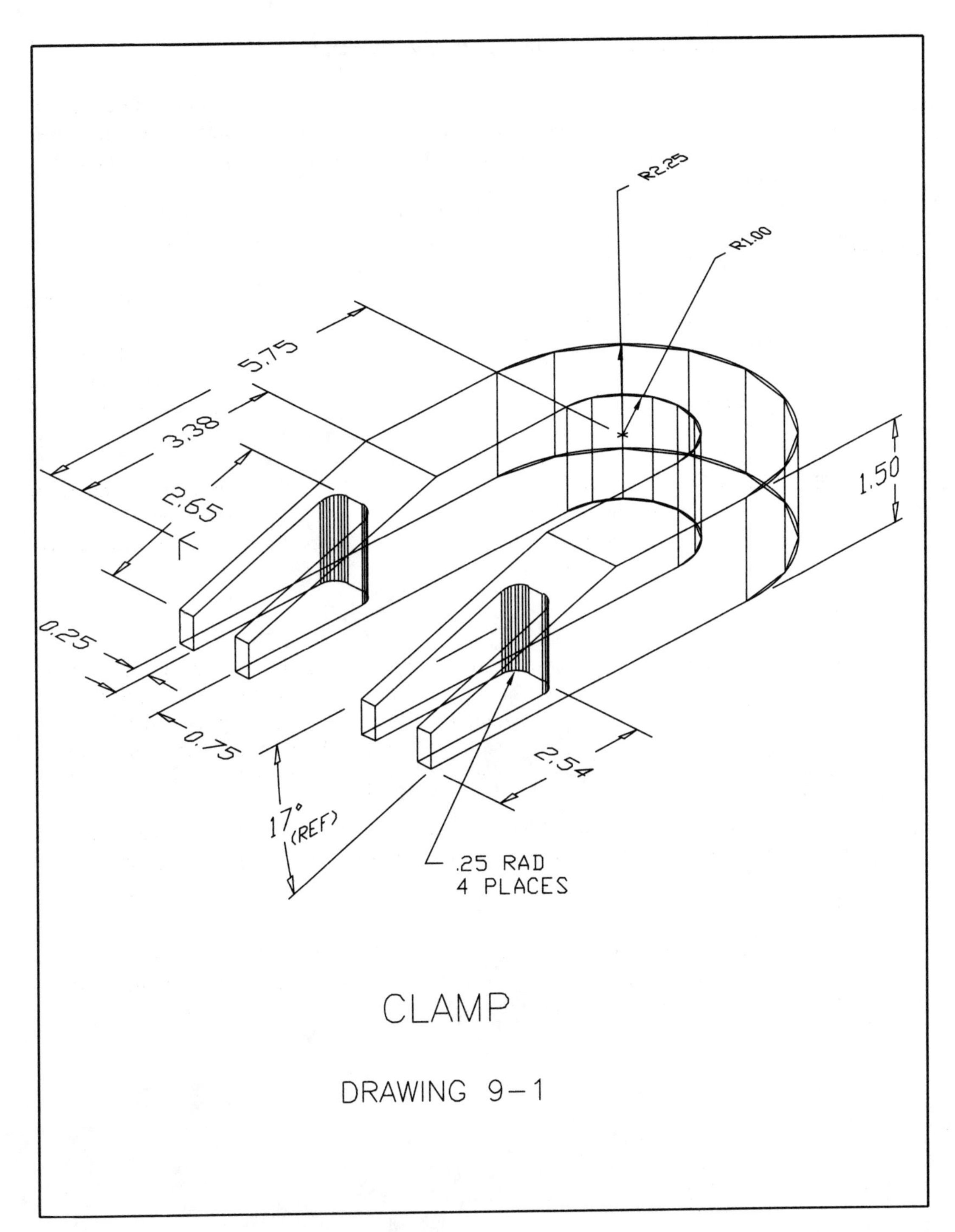
R2.25
R1.00
5.75
3.38
2.65
1.50
0.25
0.75
2.54
17°
(REF)
.25 RAD
4 PLACES
CLAMP
DRAWING 9-1

DRAWING 9-2: REVSURF DESIGNS

The REVSURF command is fascinating and powerful. As you get familiar with it, you may find yourself identifying objects in the world that can be conceived as surfaces of revolution. To encourage this process, we have provided this page of 12 REVSURF objects and designs.

To complete the exercise, you will need only the LINE and REVSURF commands. In the first six designs, we have shown the path curves and axes of rotation used to create the design. In the other six, you will be on your own.

Exact shapes and dimensions are not important in this exercise. Imagination is. When you have completed our designs, we encourage you to invent a number of your own.

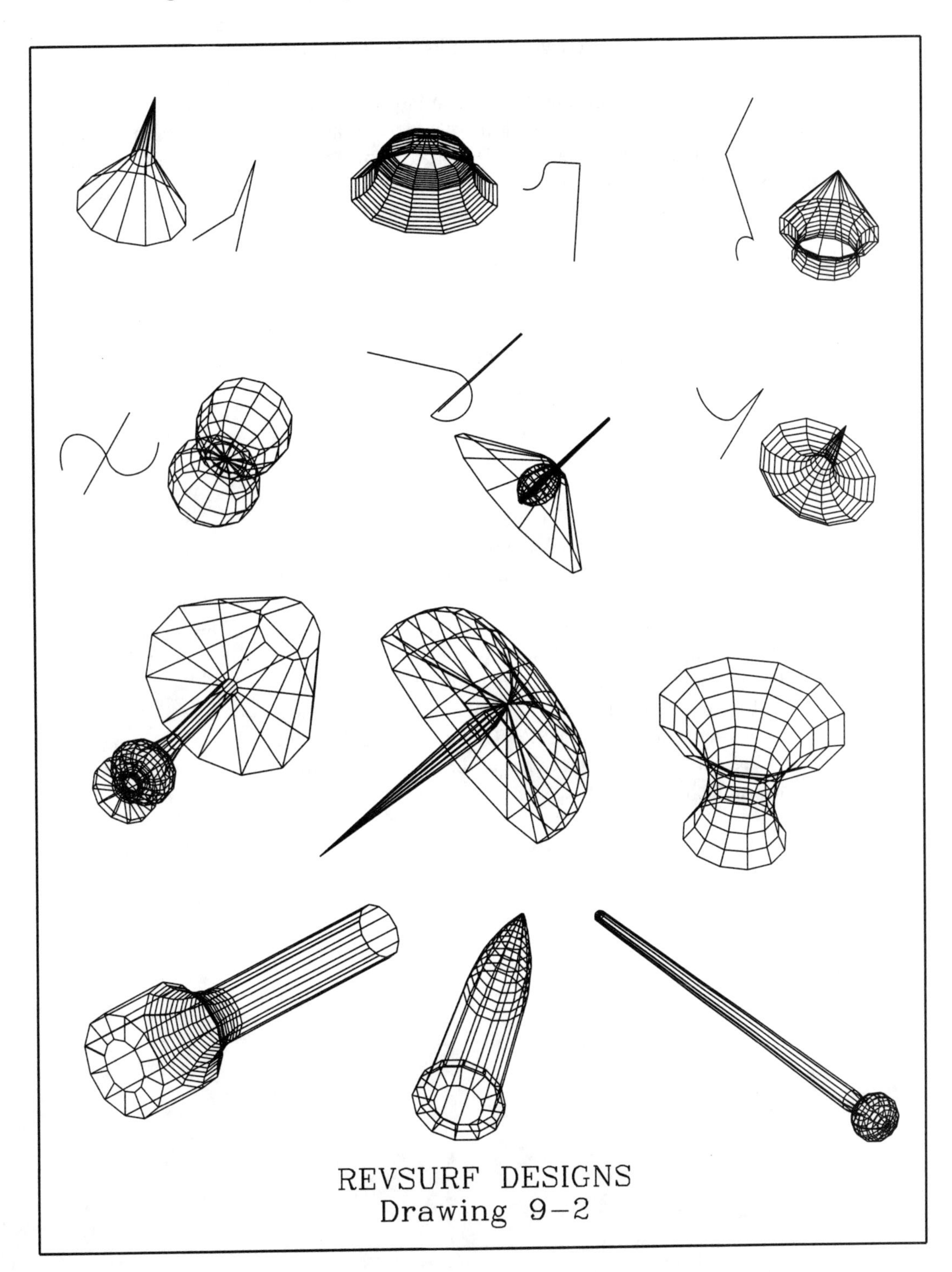
REVSURF DESIGNS
Drawing 9-2

DRAWING 9-3: BUSHING MOUNT

It is important to use an efficient sequence in the construction of composite solids. In general, this will mean saving union, subtraction, and intersection operations until most of the solid objects have been drawn and positioned. This approach has two advantages. It takes up less memory and it allows you to continue to use the geometry of the parts for snap points as you position other parts.

DRAWING SUGGESTIONS

> Use at least two views, one plan and one 3D, as you work.
> Begin with the bottom of the mount in the XY plane. This will mean drawing a 6.00 $\times$ 4.00 $\times$ .50 solid box sitting on the XY plane.
> Draw a second box, 1.50 $\times$ 4.00 $\times$ 3.50, in the XY plane. This will become the upright section at the middle of the mount. Move it so that its own midpoint is at the midpoint of the base.
> Draw a third box, 1.75 $\times$ .375 $\times$ .50, in the XY plane. This will be copied and become one of the two slots in the base. Move it so that the midpoint of its long side is at the midpoint of the short side of the base. Then MOVE it 1.25 along the X axis.
> Add a .375 radius cylinder with .50 height at each end of the slot.
> Copy the box and cylinders 3.75 to the other side of the base to form the other slot.
> Create a new UCS 2.00 up in the Z direction. You can use the origin option and give (0,0,2) as the new origin. This puts the XY plane of the UCS directly at the middle of the upright block, where you can easily draw the bushing.
> Move out to the right of the mount and draw the polyline outline of the bushing as shown in the drawing. Use REVOLVE to create the solid bushing.
> Create a cylinder in the center of the mount, where it can be subtracted to create the hole in the mount upright.
> Union the first and second boxes.
> Subtract the boxes and cylinders to form the slots in the base and the bushing-sized cylinder to form the hole in the mount.

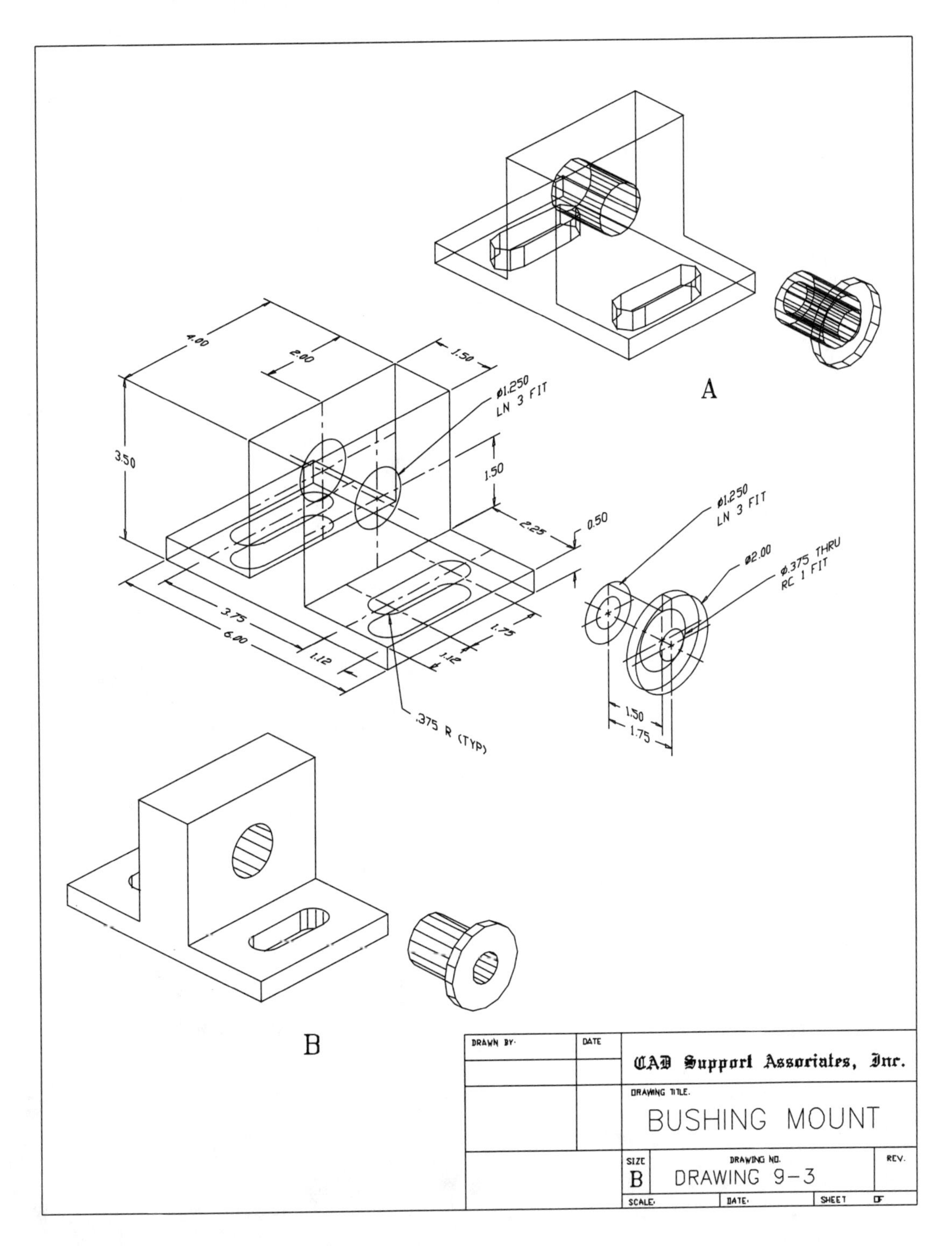
A
4.00
2.00
1.50
ø1.250
LN 3 FIT
3.50
1.50
2.25
0.50
3.75
6.00
1.12
1.75
1.12
.375 R (TYP)
ø1.250
LN 3 FIT
ø2.00
ø.375 THRU
RC 1 FIT
1.50
1.75
B
DRAWN BY:
DATE
CAD Support Associates, Inc.
DRAWING TITLE:
BUSHING MOUNT
SIZE
B
DRAWING NO.
DRAWING 9-3
REV.
SCALE:
DATE:
SHEET
OF

Index

N

O

P

Q

R

S